Lecture Notes in Physics

302

François Gieres

Geometry of Supersymmetric Gauge Theories

Including an Introduction to BRS Differential Algebras and Anomalies

Springer-Verlag
Berlin Heidelberg GmbH

Author

François Gieres
Institut für Theoretische Physik der Universität Bern
Sidlerstraße 5, CH-3012 Bern

ISBN 978-3-662-13677-5 ISBN 978-3-540-39100-5 (eBook)
DOI 10.1007/978-3-540-39100-5

Originally published by Springer-Verlag Berlin Heidelberg New York in 1988
Softcover reprint of the hardcover 1st edition 1988

2158/3140-543210

PREFACE

In this set of notes we give a systematic and essentially self-contained account of the geometric formulation of supersymmetric YM-theories (SYM-theories). In our treatment, which is primarily concerned with the classical theory, we have included a detailed discussion of the BRS differential algebras, which represent the starting point for the quantization of these theories*). While a large amount of the material covered is widely known, many topics are presented in a more systematic and geometric way than usual. A detailed and informative presentation was attempted, with the exception of topics discussed in great detail in the literature, which are only described briefly. The list of references is not intended to be exhaustive, but does include pioneering work as well as some recent reviews.

Different formulations of a general theory of supermanifolds have been proposed in recent years, but in their present form none of them is completely satisfactory from both the mathematical and physical point of view**) and therefore we have refrained from using this approach here. The superspace notations and conventions used which are summarized in Appendix 1, are those of Wess and Bagger[154)], including the appropriate $N > 1$ generalization presented in Refs.65) and 126); in particular, we have restricted ourselves to four space-time dimensions.

This book begins with a detailed account of the geometry of rigid superspace using the appropriate mathematical techniques and terminology. For example, we show that rigid superspace can be viewed as a particular reductive homogeneous space and that the quantities which are generally used to define its geometric structure (the "flat vielbein forms" and the "susy covariant derivatives") are local expressions for a canonically given linear connection. Furthermore the supersymmetry transformations are defined in a geometric way and comparison is made with the super-Riemannian approach. Next we introduce the different representations of $N = 1$ SYM-theories and study their interrelations, which are typical for constrained globally and locally supersymmetric theories. Particular attention is paid to the reality properties of the superfields, essential for obtaining the proper field theory upon projection to space-time. Following on, Part III contains a comprehensive account of the so-called real representation of SYM-theories and of its characteristic features.

There is a certain gap in the scientific literature between the usual textbook presentation of the BRS transformations and the very mathematical treatment based on differential algebras encountered in recent publications. After carefully filling

---

*) A comprehensive account of the quantum theory can be found in the textbooks of Gates et al.[57)] and of Piquet and Sibold[114)].

**) See e.g., Refs.86) and 117) and the remarks made in Sect.I.1.

this gap, we consider the BRS differential algebras and the possible anomalous terms of SYM-theories. This part closes with a systematic study of the BRS algebra in the WZ-gauge, which represents a prototype for similar, although technically more complicated, algebras occurring in supergravity and superstring theories.

The final chapters are devoted to a generalization of the preceding work to extended SYM-theories. As far as the general geometric set-up is concerned, this can be done in a straightforward way, but the formulation and the solution of the constraints on the superconnection lead to complications, which are discussed.

At the end of the book we have included several appendices on supersymmetry and on the so-called BRS-cohomology. The last appendix, which is based on some unpublished notes of R. Stora, provides a self-contained account of anticommuting spinors in ordinary and supersymmetric field theories.

While preparing these notes, I have benefited from many pleasant and stimulating discussions with experts in the field. I am particularly indebted to F. Delduc, G. Girardi, R. Grimm and R. Stora for their valuable advice, for their patience and encouragement, as well as for their critical remarks on the manuscript. For helpful conversations and comments, I would like to thank L. Bonora, M. Forger, M.T. Grisaru, J. McCabe, O. Piguet, P. Sorba, J.C. Wallet and J. Wess. I express my gratitude to P. Sorba and all the members of the theory group of LAPP (Annecy-le-Vieux) for their warm welcome and for the pleasant atmosphere. The last part of the text was prepared at the Institute for Theoretical Physics of the University of Berne, and I thank Professor H. Leutwyler and all the members of the Institute for their kind hospitality. Financial support from Ministère de la Recherche et de l'Enseignement Supérieur, France, from Ministère de l'Education Nationale et de la Jeunesse, Luxembourg, and from the Eidgenössische Stipendienkommission, Switzerland, is thankfully acknowledged. The manuscript was typed by N. Berger and O. Hänni, to whom I am indebted for their neat and skilful work.

Bern, March 1988

François Gieres

TABLE OF CONTENTS

I. THE CANONICAL GEOMETRIC STRUCTURE OF RIGID SUPERSPACE AND SUSY TRANSFORMATIONS

I.1 Introduction to part I

The concept of superspace introduced in 1974 by A. Salam and J. Strathdee[118)] was soon realized to represent the appropriate device for the formulation of supersymmetric field theories. Besides its applications in Modern Particle Physics it has stimulated the research in various subfields of Mathematical Physics and Pure Mathematics. Nowadays many fine reviews of the basic concepts and techniques are available in the literature[57,128,154,155)], but unfortunately a certain mathematical terminology which is not always very clear and meaningful became quite popular in this field. (Some examples we have in mind are the "susy covariant derivatives", the "flat vielbein forms" and the "flat superspace torsion".) In the following chapters we will review the geometry of rigid superspace and try to clarify the basic concepts and their geometric interpretation[*)].

Here, as well as in the other parts of these notes, our main concern will be the geometric framework and not the analytic one and our attitude towards the superdifferentiable calculus will be the following one. At this moment a completely satisfying theory of supergroups and manifolds with an appropriate differential and integral calculus (involving superfunctions with anticommuting component fields) is still lacking. (Some shortcomings of different approaches to such a theory have been pointed out recently[86,117)]; for detailed reviews and further references, see 42),54),88),122)). Also at a basic conceptual level the exact relation between the very general mathematical work and the heuristic approach considered in the formulation of physical theories is not completely clear and still a matter of current research[23)]. For these reasons we shall adopt the pragmatic view-point which is usually taken in this context, i.e. we ignore all questions of differential structure and restrict ourselves to local considerations, hoping that the missing mathematical justifications will be supplied later on.

However we would like to stress that the basic quantity in the geometric approach to supersymmetry is not the (rigid or flexible) superspace itself, but the algebra of superfunctions, i.e. superfields. The supersymmetry generators and covariant derivatives represent derivations on this graded commutative algebra whereas the vielbein forms describe linear forms on the set of derivations. Accordingly the sheaf-theoretic approach which starts with an ordinary space-time manifold and extends the commutative algebra of ordinary functions to a graded commutative one, is best suited for the needs of Physics. These mathematical ideas which go back to J. von Neumann and his contemporaries have been worked out for the

---

*) We will generally use the terms "rigid" and "flexible" superspace[163)] instead of the familiar ones "flat" and "curved" space for reasons which will become apparent in chapter I.9.

present case by B. Kostant[88)] and M. Batchelor[179)]. For a nice introduction to these topics and their application to supersymmetry we refer to an article by J. Dell and L. Smolin[180)]. (Unfortunately these authors didn't take into account the anticommuting character of the fermionic fields involved in the superfield expansions. But this problem which is not characteristic for supersymmetry can easily be overcome by a generalization of Schwinger's treatment of spinor fields[*)].

In view of the applications to be considered in the following parts of the text we concentrate on the N=1, d=4 rigid superspace, although the restriction to four space-time dimensions is not really essential in this context and was only chosen for notational reasons.

I.2 Definition of rigid superspace

Usually rigid superspace is viewed as a certain set of points $(x,\theta,\bar{\theta})$ which is equipped, not only with an appropriate differentiable structure, but also with an additional structure specified by certain vielbein forms $E^A = dz^M E_M{}^A$. In the following we will formulate this definition in a mathematically more precise and standard way.

(i) General definition

Following the usual approach we start with the N=1, d=4 super-Poincaré algebra (without central extension) which we denote by Lie SP; its generators M_{mn}, P_m, Q_α, $\bar{Q}_{\dot{\alpha}} = (Q_\alpha)^*$ satisfy the commutation relations:

$$\begin{aligned}
&[M_{mn}, M_{lp}] = i(\eta_{nl} M_{mp} - \eta_{np} M_{ml} - \eta_{ml} M_{np} + \eta_{mp} M_{nl}) \\
&[P_m, M_{nl}] = i(\eta_{mn} P_l - \eta_{ml} P_n) \\
&[Q_\alpha, M_{mn}] = \frac{1}{2} (\sigma_{mn})_\alpha{}^\beta Q_\beta \\
&[\bar{Q}_{\dot{\alpha}}, M_{mn}] = -\frac{1}{2} \bar{Q}_{\dot{\beta}} (\bar{\sigma}_{mn})^{\dot{\beta}}{}_{\dot{\alpha}} \\
&\{Q_\alpha, \bar{Q}_{\dot{\beta}}\} = 2\sigma^m_{\alpha\dot{\beta}} P_m \\
&0 = \{Q_\alpha, Q_\beta\} = \{\bar{Q}_{\dot{\alpha}}, \bar{Q}_{\dot{\beta}}\} = [P_m, Q_\alpha] = [P_m, \bar{Q}_{\dot{\alpha}}] = [P_m, P_n] \;.
\end{aligned} \tag{1.1}$$

*) See appendix A.6.

This super Lie algebra*) has a very particular structure for it is the semi-direct sum of two algebras, the Lie algebra of the Lorentz group (Lie L, with generators M_{mn}) and the (graded) supersymmetry algebra ($\mathcal{M}$, with generators P_m, Q_α, $\bar{Q}_{\dot{\alpha}}$) :

$$\mathcal{L}ie\ SP = \mathcal{L}ie\ L \oplus \mathcal{M} \tag{1.2}$$

[Lie L, Lie L] $\subset$ Lie L, $[\mathcal{M},\mathcal{M}] \subset \mathcal{M}$, [Lie L, $\mathcal{M}$] $\subset \mathcal{M}$

("The Lorentz transformations act on the susy algebra".)

Using constant anticommuting Weyl spinors $\theta_\alpha, \bar{\theta}_{\dot{\alpha}} = (\theta_\alpha)^*$ and real parameters x^m, $\lambda^{mn} = -\lambda^{nm}$ the graded algebra Lie SP can be formally exponentiated to yield the super-Poincaré group SP parametrized by:

$$G(x,\theta,\bar{\theta};\lambda) = \exp(-ix^m P_m + i\theta^\alpha Q_\alpha + i\bar{\theta}_{\dot{\alpha}}\bar{Q}^{\dot{\alpha}})\exp\left(\frac{i}{2}\lambda^{mn}M_{mn}\right). \tag{1.3}$$

By definition rigid superspace is the quotient space

$$M = SP/L$$

and thus may be parametrized by local coordinates $(z^M) = (x^m,\theta^\alpha,\bar{\theta}_{\dot{\alpha}})$. It is to be considered as a generalization of Minkowski space which can be defined (from the set-theoretical point of view) as P/L ; while the algebra of ordinary translations is abelian, the supersymmetry algebra is not and therefore rigid superspace SP/L is expected to "have a richer structure" than Minkowski space.

Concerning the possible differential structures on quotient spaces we mention the following basic result from the bosonic case[150]. If G is a Lie group and H a closed subgroup of G, then G/H has a unique manifold structure such that the natural projection map $\pi : G \to G/H$ is smooth and such that there exist local smooth sections of G/H in G ; the quotient space G/H endowed with this differential structure is then referred to as a homogeneous manifold.

(ii) Rigid superspace as a reductive homogeneous space

When we discuss invariant linear connections on rigid superspace, we will rely on a slightly different definition of this space, namely

$$M = SP_0/L_0$$

*) For a formal definition of super vector spaces and Lie algebras we refer to the textbooks of M. Scheunert[191] and B. de Witt[42] and to the overview article of V.G. Kac[181].

where SP_o and L_o denote the connected components of the identity of the respective groups. This definition is essentially equivalent to the first one since the four connected components of SP exactly correspond to those of L.

The group L_o being connected, the property $[\mathrm{Lie}\ L_o, \mathfrak{M}] \subset \mathfrak{M}$ is equivalent to

$$\mathrm{ad}(L_o)\,\mathfrak{m} \subset \mathfrak{m}$$

where $\mathrm{ad}(L_o)$ denotes the action of the adjoint representation of the group L_o on the susy algebra $\mathfrak{M}$*). The last relation implies that $M = SP_o/L_o$ represents (the graded generalization of) a reductive homogeneous space. We recall[85] that a homogeneous space $M = G/H$, with G a connected Lie group and H a closed subgroup, is said to be reductive, if the Lie algebra $\mathfrak{g}$ of G can be decomposed into a vector space direct sum of the Lie algebra h of H and an $\mathrm{ad}(H)$-invariant subspace $\mathfrak{M}$, i.e. if

$$\mathfrak{g} = h + \mathfrak{M}$$

with

$$h \cap \mathfrak{M} = \{0\}$$

and

$$\mathrm{ad}(H)\mathfrak{M} \subset \mathfrak{M}.$$

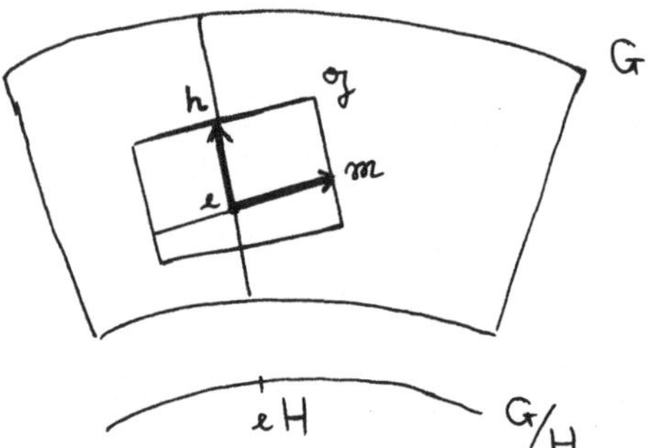

Note that SP_o/L_o represents a very particular reductive homogeneous space, because in this case the vector subspace $\mathfrak{M}$ actually represents a (graded) Lie algebra.

As a consequence of the general mathematical theory the space SP_o/L_o admits a canonical SP_o-invariant linear connection which will be discussed in chapter I.9; we will see that SP_o/L_o supplemented with this connection corresponds to the usual definition of "flat superspace" in terms of certain vielbein forms. The corresponding curvature form vanishes, but the torsion does not; therefore rigid superspace equipped with the canonical connection is not a flat space in the usual mathematical sense (sect. I.9 (iii)).

We note that reductive homogeneous spaces have a wide range of applications in Physics and occur for instance in the study of σ-models[109], YM-theories[7], (super-) gravity[102] and Kaluza-Klein theories[146].

---

*) For a proof of this equivalence consider for instance the BCH-formula in the form
$$e^{A}\, B\, e^{-A} = B + [A,B] + \ldots$$

(iii) The Riemannian view-point

In the literature rigid superspace defined as $M = SP/L$ was occasionally supplemented with a supermetric (g_{MN}) and thus turned into a super-Riemannian space (with non-vanishing Riemannian curvature)[159,129,42]. Although we do not think that this is the adequate view-point, it is interesting to compare this approach to the familiar one; this will be done in chapter I.10.

I.3 The left action on superspace

(i) The method of induced representations ; definition of susy transformations

The group elements $G(a,\zeta,\bar{\zeta}\ ;\ \varepsilon=0)$ obtained by exponentiation of the susy algebra define a subgroup of the super-Poincaré group SP. Their left action on SP,

$$\ell_{(a,\zeta,\bar{\zeta};0)}\, G(x,\theta,\bar{\theta};\lambda) \equiv G(a,\zeta,\bar{\zeta};0)\cdot G(x,\theta,\bar{\theta};\lambda)$$

can easily be evaluated by use of the BCH-formula $e^A e^B = e^{A+B+\frac{1}{2}[A,B]+\dots}$ and the SP-algebra relations (1.1) : the induced action on the quotient space $M = SP/L$ reads

$$f : M \longrightarrow M \tag{1.4}$$

$$z = (x^m, \theta^\alpha, \bar{\theta}_{\dot{\alpha}}) \longmapsto z' = f(z) = \left(x^m + a^m + i(\theta\sigma^m\bar{\zeta} - \zeta\sigma^m\bar{\theta}),\ \theta^\alpha + \zeta^\alpha,\ \bar{\theta}_{\dot{\alpha}} + \bar{\zeta}_{\dot{\alpha}}\right).$$

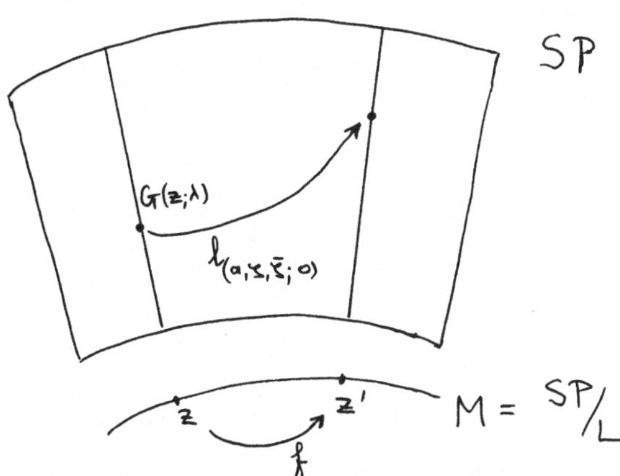

Due to the particular form of the algebra (1.1) these transformations do not only hold to first order in a^m and ζ_α (i.e. for a^m and ζ_α small w.r.t. an appropriate topology), but to arbitrary order.

By definition these coordinate transformations of $M = SP/L$ are the rigid susy transformations (and the ordinary translations). They can be interpreted as the infinitesimal coordinate transformations $z'^M = z^M + \xi^M$ generated by the super vector field

$$\xi = \left(a^m + i\theta\sigma^m\bar{\zeta} - i\zeta\sigma^m\bar{\theta}\right)\frac{\partial}{\partial x^m} + \zeta^\mu \frac{\partial}{\partial\theta^\mu} + \bar{\zeta}_{\dot{\mu}}\frac{\partial}{\partial\bar{\theta}_{\dot{\mu}}} \tag{1.5}$$

Here $(\partial_M) \equiv (\partial/\partial x^m, \partial/\partial\theta^\mu, \partial/\partial\bar{\theta}_{\dot{\mu}})$ represents the natural, i.e. coordinate-induced basis of the super-tangent space T_zM *).

To determine the basis of T_zM which is induced by the left action of the group element $G(a,\zeta,\bar{\zeta}\ ;\ 0)$, we rewrite the vector field (1.5) in the form

$$\xi = a^a \hat{Q}_a + \zeta^\alpha \hat{Q}_\alpha + \bar{\zeta}_{\dot{\alpha}} \hat{\bar{Q}}^{\dot{\alpha}}$$

Comparison with (1.5) yields

$$\hat{Q}_a = \delta_a^{\ m}\partial_m \ , \quad \hat{Q}_\alpha = \frac{\partial}{\partial\theta^\alpha} - i\sigma^m_{\alpha\dot{\beta}}\bar{\theta}^{\dot{\beta}}\partial_m \ , \quad \hat{\bar{Q}}^{\dot{\alpha}} = \frac{\partial}{\partial\bar{\theta}_{\dot{\alpha}}} - i\theta^\alpha\sigma^m_{\alpha\dot{\beta}}\varepsilon^{\dot{\beta}\dot{\alpha}}\partial_m \ . \tag{1.6}$$

Since $\hat{Q}_A = \tilde{E}_A^{\ M}\partial_M$ with $(\tilde{E}_A^{\ M}(z))$ representing an invertible matrix, the first order differential operators $(\hat{Q}_A) = (\hat{Q}_a, \hat{Q}_\alpha, \hat{\bar{Q}}^{\dot{\alpha}})$ actually define a basis of T_zM. In the following we will mainly consider these differential operators and not the susy algebra generators P_m, Q_α, $\bar{Q}^{\dot{\alpha}}$ and thus no confusion should arise from suppressing the hats on the expressions (1.6).

For any two vector fields $\xi_1 = \xi_1^M(z)\partial_M$ and $\xi_2 = \xi_2^M(z)\partial_M$ one can define the Lie bracket of ξ_1 and ξ_2: this is again a vector field denoted by $[\xi_1,\xi_2]$ and given by

$$[\xi_1 , \xi_2] = [\xi_1^N(\partial_N\xi_2^M) - \xi_2^N(\partial_N\xi_1^M)]\,\partial_M \ . \tag{1.7}$$

As may be directly verified, all the Lie brackets between the tangent vectors Q_a, Q_α, $\bar{Q}^{\dot{\alpha}}$ vanish except the (graded) bracket

$$[Q_\alpha , \bar{Q}_{\dot{\beta}}] = 2i\,\sigma^c_{\alpha\dot{\beta}}\,Q_c \ .$$

We can rephrase this result as follows. While the tangent vectors (∂_M) represent a holonomic basis of T_zM,

$$[\partial_M , \partial_N] = 0 \ ,$$

*) Since most geometric objects considered in the body of these notes live on rigid superspace, we will often suppress the prefix "super" and simply talk about vector fields, p-forms,...

the vectors (Q_A) define an anholonomic basis which is characterized by the constant*) anholonomy-coefficients $2i\sigma^c_{\alpha\dot\beta}$.

Alternatively the differential operators $(-i\, Q_A)$ can be regarded as defining a linear representation of the susy algebra on the vector space of superfields, i.e. superfunctions on M :

$$F(z^M + \xi^M) = F(z) + \delta_\xi F + \sigma(\xi^2)$$

with

$$\delta_\xi F = \xi^M \partial_M F = (a^m \partial_m + \xi^\alpha Q_\alpha + \bar\xi_{\dot\alpha} \bar Q^{\dot\alpha}) F$$

$$[-i\, Q_\alpha, -i\, \bar Q_{\dot\beta}] = 2\sigma^m_{\alpha\dot\beta}(-i\,\partial_m) \qquad (P_m \equiv -i\,\partial_m)\,. \tag{1.8}$$

Concerning the reality properties of the operators Q_A we remark the following. According to the rules for complex conjugation outlined in appendix 2 we have

$$(\partial_m)^* = \partial_m \quad , \quad (Q_\alpha)^* = \bar Q_{\dot\alpha}\ .$$

Thus the supersymmetry generators $(-iQ_\alpha)$ and $(-i\bar Q_{\dot\alpha})$ can also be thought of as being hermitean conjugates of each other w.r.t. the inner product

$$\langle F, G \rangle = \int d^8 z \; F^*(z) \cdot G(z) \quad ,$$

i.e. explicitly:

$$(-i\, Q_\alpha)^\dagger = -i\, \bar Q_{\dot\alpha} \quad \text{and} \quad (-i\,\partial_m)^\dagger = -i\,\partial_m\ .$$

Here the integration over the Grassmann variables θ^α, $\bar\theta_{\dot\alpha}$ is to be understood as a Berezinian integral[162,154] for which integration and differentiation are formally the same:

$$\int d^8 z\; F(z) \equiv \int d^4x \int d^2\theta \int d^2\bar\theta\; F(x, \theta, \bar\theta)$$

with

*) Note that for a generic basis $E_A{}^M(z)\partial_M$ of T_zM, these coefficients are in general not constant functions on M.

$$\int d^2\theta\, F(z) \equiv \frac{1}{4}\, \varepsilon^{\alpha\beta} \int d\theta_\alpha \int d\theta_\beta\, F(z) = \frac{1}{4}\, \varepsilon^{\alpha\beta} \frac{\partial}{\partial\theta^\alpha} \frac{\partial}{\partial\theta^\beta}\, F(z)$$

and

$$\int d^2\bar\theta\, F(z) \equiv \frac{1}{4}\, \varepsilon_{\dot\alpha\dot\beta} \int d\bar\theta^{\dot\alpha} \int d\bar\theta^{\dot\beta}\, F(z) = \frac{1}{4}\, \varepsilon_{\dot\alpha\dot\beta} \frac{\partial}{\partial\bar\theta_{\dot\alpha}} \frac{\partial}{\partial\bar\theta_{\dot\beta}}\, F(z)$$

(ii) <u>Action of the Lorentz group on rigid superspace</u>

The left action of the Lorentz group L on the super-Poincaré group induces an action of L on rigid superspace M = SP/L ; it can be determined from the equation[128)]

$$e^{\frac{i}{2}\lambda M}\, G(x,\theta,\bar\theta;0) =: G(x',\theta',\bar\theta';0)\, e^{\frac{i}{2}\lambda M}$$

which leads to the following transformations:

$$\begin{aligned} x'^m &= (e^{-\lambda})^m{}_n\, x^n \qquad (\lambda^{mn} = -\lambda^{nm}) \\ \theta'^\alpha &= \theta^\beta (A^{-1})_\beta{}^\alpha \equiv \theta^\beta \left[\exp(-\tfrac{i}{4}\lambda^{mn}\sigma_{mn})\right]_\beta{}^\alpha \\ \bar\theta'_{\dot\alpha} &= \bar\theta_{\dot\beta} (A^+)^{\dot\beta}{}_{\dot\alpha} \equiv \bar\theta_{\dot\beta}\left[\exp(-\tfrac{i}{4}\lambda^{mn}\bar\sigma_{mn})\right]^{\dot\beta}{}_{\dot\alpha} \end{aligned} \tag{1.9}$$

Here the matrix $A \equiv \exp(\frac{i}{4}\lambda^{mn}\sigma_{mn})$ is an element of SL(2,ℂ), since σ_{mn} is antisymmetric and thus traceless $(\det e^\sigma = e^{\mathrm{tr}\sigma} = 1)$.

I.4 Left-invariant vector fields on superspace

(i) The right action

Since the rigid susy transformations are induced from a left action on the group SP, the natural way for obtaining a theory on $M = SP/L$ which is invariant under these transformations is to rely on the fact that the left and the right action on a group commute with each other. Thus one expresses all geometrical quantities on M with respect to the basis (D_A) of T_zM (or with respect to its dual basis (E^A)) which is induced by the right action of $G(a,\zeta,\bar{\zeta};\,0)$ on SP. By the previously described method of induced representations one finds:

$$D_a = \delta_a^{\ m}\,\partial_m\ ,\quad D_\alpha = \frac{\partial}{\partial\theta^\alpha} + i(\sigma^m\bar{\theta})_\alpha\,\partial_m\ ,\quad \bar{D}^{\dot{\alpha}} = \frac{\partial}{\partial\bar{\theta}_{\dot{\alpha}}} + i\,\theta^\alpha\sigma^m_{\alpha\dot{\beta}}\,\varepsilon^{\dot{\beta}\dot{\alpha}}\,\partial_m$$

$$D_A = E_A^{\ M}\,\partial_M \quad \text{with} \quad (E_A^{\ M}) \quad \text{invertible} \tag{1.10}$$

$$[D_\alpha\,,\,\bar{D}_{\dot{\beta}}] = -2i\,\sigma^c_{\alpha\dot{\beta}}\,D_c$$

the explicit expression for $(E_A^{\ M})$ being

$$(E_A^{\ M}) = \begin{bmatrix} E_a^{\ m} = \delta_a^{\ m} & E_a^{\ \mu} = 0 & E_{a\dot{\mu}} = 0 \\ E_\alpha^{\ m} = i(\sigma^m\bar{\theta})_\alpha & E_\alpha^{\ \mu} = \delta_\alpha^{\ \mu} & E_{\alpha\dot{\mu}} = 0 \\ E^{\dot{\alpha}m} = i(\bar{\sigma}^m\theta)^{\dot{\alpha}} & E^{\dot{\alpha}\mu} = 0 & E^{\dot{\alpha}}_{\ \dot{\mu}} = \delta^{\dot{\alpha}}_{\ \dot{\mu}} \end{bmatrix}. \tag{1.11}$$

The differential operators $(-iD_A)$ represent the susy algebra up to a sign,

$$[-i\,D_\alpha\,,\,-i\,\bar{D}_{\dot{\beta}}] = -2\,\sigma^m_{\alpha\dot{\beta}}\,(-i\,\partial_m) \tag{1.12}$$

and satisfy the reality conditions

$$(D_\alpha)^* = \bar{D}_{\dot{\alpha}}$$

resp.

$$(-i\,D_\alpha)^\dagger = -i\,\bar{D}_{\dot{\alpha}}$$

The different signs in the graded commutators $[-iQ_\alpha, -i\bar{Q}_{\dot\beta}]$ and $[-iD_\alpha, -i\bar{D}_{\dot\beta}]$ reflect the well-known fact*) that the left action of a Lie group on a manifold M induces a homomorphism of Lie algebras,

$$\text{Lie } G \longrightarrow \{\text{Vector fields on } M\}$$
$$X \longrightarrow \hat{X} \text{ with } \widehat{[X,Y]} = [\hat{X},\hat{Y}],$$

whereas the right action induces only a homomorphism up to a sign, $\widehat{[X,Y]} = -[\hat{X},\hat{Y}]$. (Note that without the factors (-i) the terms right and left have to be exchanged in this statement.)

(ii) <u>Invariance and covariance properties of (D_A)</u>

The commutativity of the left and right actions on the group SP is expressed on the quotient space M by the fact that the vector fields (D_A) and (Q_A) have a vanishing Lie bracket :

$$[Q_B, D_A] = 0$$

resp.

$$L_\xi D_A = [\xi, D_A] = [a^b Q_b + \zeta^\beta Q_\beta + \bar{\zeta}_{\dot\beta} \bar{Q}^{\dot\beta}, D_A] = 0 \,. \qquad (1.13)$$

The <u>invariance of the vector field (D_A) under the susy transformations (1.4)</u> can also be described as follows: replace x, θ and $\bar\theta$ by x', θ' and $\bar\theta'$, then $D'_A = D_A$ or in more mathematical terms,

$$f_* D_A = D_A \qquad (1.14)$$

where $f(z) = z'$ represents the susy transformations and where f_* denotes the projection ("push-forward") of tangent vectors induced by f [36]:

$$f_* : T_z M \longrightarrow T_{z'} M \,.$$

In the literature the vectors (D_A) are usually referred to as "susy covariant derivatives". In section I.9 (ii) we will see that they can indeed be considered as covariant derivatives w.r.t. the canonical linear connection on SP_o/L_o, but for the time being and in view of the equations (1.13),(1.14) we will refer to (D_A) as the <u>(left-) invariant basis of T_zM.</u>

---

*) See e.g. ref. 85), Errata to p.106.

Of course the vectors (D_A) transform covariantly under the Lorentz transformations (1.9):

$$D_a' = (e^{\lambda})^b{}_a D_b = (e^{-\lambda})_a{}^b D_b$$
$$D_\alpha' = A_\alpha{}^\beta D_\beta \qquad (1.15)$$
$$\bar{D}^{\dot\alpha\prime} = (A^{-1\dagger})^{\dot\alpha}{}_{\dot\beta} \bar{D}^{\dot\beta}$$

(iii) <u>Particularities of the anholonomic basis (D_A)</u>

There is a price to pay for using the invariant basis (D_A) of T_zM : its <u>anholonomy coefficients $\mathcal{C}_{AB}{}^C$</u> which are given by

$$[D_A , D_B] = \mathcal{C}_{AB}{}^C D_C \qquad (1.16)$$
$$\mathcal{C}_{\alpha\dot\beta}{}^c = -2i\,\sigma^c_{\alpha\dot\beta} = \mathcal{C}_{\dot\beta\alpha}{}^c \qquad \text{(other components = 0)}$$

show up in the Lie bracket of any two vector fields if these are expressed with respect to the basis (D_A) :

$$[\xi_1 , \xi_2] = [\xi_1^A (D_A \xi_2^B) - \xi_2^A (D_A \xi_1^B)] D_B + \xi_1^A \xi_2^B \mathcal{C}_{BA}{}^C D_C .$$

As an illustrative example one might consider the susy generating vector field (1.5): under the change of basis $\partial_M \to D_A = E_A{}^M \partial_M$ its components become $\xi^A = \xi^M E_M{}^A$, i.e. explicitly:

$$(\xi^A) = \left(\delta_a{}^m (a^m + 2i\theta\sigma^m\bar\xi - 2i\xi\sigma^m\bar\theta),\ \delta^\alpha{}_\mu \xi^\mu ,\ \delta_{\dot\alpha}{}^{\dot\mu} \bar\xi_{\dot\mu}\right). \qquad (1.17)$$

Notice the factor two in ξ^a which is not present in ξ^m and which requires some care in the choice of indices in this context. (Of course the vector field ξ itself is invariant under any change of basis, in particular $\xi^M\partial_M = \xi^A D_A$.)

I.5 Left-invariant forms on superspace

(i) The dual basis (E^A)

For an ordinary manifold the dual basis (dx^m) of the basis (∂_m) of T_xM is defined by the relation

$$dx^m(\partial_n) = \delta^m{}_n$$

and the exterior product of two monomials dx^m, dx^n satisfies the anti-commutative law

$$dx^m \wedge dx^n = -dx^n \wedge dx^m .$$

Furthermore the exterior differential d acts on p-forms $\alpha = \frac{1}{p!}\alpha_{i_1\ldots i_p} dx^{i_1} \wedge \ldots \wedge dx^{i_p}$ (with $\alpha_{i_1\ldots i_p}$ totally antisymmetric) as

$$d\alpha = \frac{1}{p!}\partial_m \alpha_{i_1\ldots i_p} dx^m \wedge dx^{i_1} \wedge \ldots \wedge dx^{i_p}$$

and it satisfies the anti-Leibniz rule

$$d(\alpha \wedge \beta) = d\alpha \wedge \beta + (-)^p \alpha \wedge d\beta \quad (\alpha = \text{p-form}).$$

For rigid superspace the dual basis (dz^M) of the basis (∂_M) of T_zM can be defined in an analogous manner:

$$dz^M(\partial_N) = \delta^M{}_N .$$

It is convenient to assign a bidegree (p,q) to the differentials dz^M : p = differential form degree, q = statistics degree ("Grassmann parity" : q = 0 for bosonic and q = 1 for fermionic variables). Thus the exterior product of two monomials dz^M, dz^N is graded anti-commutative*):

$$dz^M \wedge dz^N = -(-)^{m\cdot n} dz^N \wedge dz^M . \tag{1.18}$$

A p-superform is most conveniently written as

$$Q = \frac{1}{p!} dz^{M_1} \ldots dz^{M_p} Q_{M_p \ldots M_1} \tag{1.19}$$

in which expression one always has an even number of indices between those being

*) E.g. $dx^m \wedge dx^n = -dx^n \wedge dx^m$, $dx^m \wedge d\theta^\alpha = -d\theta^\alpha \wedge dx^m$, $d\theta^\alpha \wedge d\theta^\beta = d\theta^\beta \wedge d\theta^\alpha$.

summed over. (Here and in the following we have suppressed the exterior product symbol $\wedge$.) The exterior differentiation is then defined by

$$dQ = \frac{1}{p!} dz^{M_1} \ldots dz^{M_p} dz^N \partial_N Q_{M_p \ldots M_1} \tag{1.20}$$

which implies

$$d(PQ) = P\, dQ + (-)^q (dP)\, Q \qquad (Q = q\text{-form}). \tag{1.21}$$

Note that this operation corresponds to a differentiation from the right. Furthermore we remark that in this set-up the statistics degree has only a passive book-keeping rôle and is not sensitive to the operator d*). At first sight these unusual rules do not seem to be very satisfying, but closer inspection shows that they describe the appropriate and mnemonically best suited superspace generalization of the ordinary differential calculus : apart from the fact that the "left-differentiation" becomes a "right-differentiation" the general rules remain the same, there are no extra sign factors and the bosonic components keep their usual form.

In view of these conventions the basis $E^A = dz^M E_M{}^A$ which represents the dual of the basis $D_A = E_A{}^M \partial_M$ of $T_z M$ has to be defined by the equation

$$D_A (\overleftarrow{E}^B) = \delta_A{}^B$$

or by the equivalent and much more convenient relation

$$d = dz^M \partial_M =: E^A D_A \quad . \tag{1.22}$$

Its explicit form is given by:

$$\begin{aligned} E^a &= dx^a - i\,(d\theta)\sigma^a \bar{\theta} + i\,\theta\sigma^a (d\bar{\theta}) \\ E^\alpha &= d\theta^\alpha \\ E_{\dot{\alpha}} &= d\bar{\theta}_{\dot{\alpha}} \quad . \end{aligned} \tag{1.23}$$

---

*) Actually one can obtain an equivalent formulation of the theory by considering a simple degree which is the sum of the form and the statistics degree[163].

(ii) <u>Invariance properties of (E^A) : geometric characterization of susy transformations</u>

Like the basis (D_A) its dual basis <u>(E^A) is invariant under the rigid susy transformations (1.4)</u>:

$$E'^A = E^A \qquad \text{i.e.} \qquad f^* E^A = E^A \tag{1.24}$$

where $f^* : T^*_{z'}M \to T^*_z M$ denotes the pull back of super 1-forms induced by the map $f : z \mapsto z'$.

In the literature the 1-forms E^A are usually called the "flat superspace vielbeins", but in view of the relation (1.24) we will refer to them as the <u>(left-) invariant basis of $T^*_z M$</u> or the <u>(left-) invariant local frames</u>.

As for the tangent vectors (D_A) there is another way for expressing the invariance of E^A. In fact the change of the 1-forms E^A under an arbitrary coordinate transformation generated by a supervector field ξ can be described by the Lie derivative[36]:

$$L_\xi E^A \equiv (i_\xi d + d i_\xi) E^A \quad . \tag{1.25a}$$

Here i_ξ denotes the interior product with respect to ξ. It acts on the algebra of superforms as an anti-derivation of degree -1,

$$i_\xi (PQ) = P(i_\xi Q) + (-)^q (i_\xi P) Q \qquad (Q = q\text{-form})$$

and it is defined on 0-forms (i.e. superfields) and 1-forms by

$$i_\xi \Phi = 0 \quad , \quad i_\xi dz^M = \xi^M \, .$$

Thus

$$i_\xi E^A = i_\xi (dz^M E_M{}^A) = \xi^M E_M{}^A$$

and more generally

$$i_\xi \left(\frac{1}{p!} dz^{M_1} \ldots dz^{M_p} Q_{M_p \ldots M_1} \right) = \frac{1}{(p-1)!} dz^{M_1} \ldots dz^{M_{p-1}} \xi^{M_p} Q_{M_p M_{p-1} \ldots M_1} \, .$$

In the evaluation of the expression (1.25a) one has to take into account the anti-commutative law (1.18) which implies

$$dE^A = \frac{1}{2} dz^M dz^N \left(\partial_N E_M{}^A - (-)^{nm} \partial_M E_N{}^A \right) .$$

The various sign factors occuring in $L_\xi E^A$ cancel each other and the final equation has the same form as in ordinary space:

$$L_\xi E^A = dz^M \left[\xi^N (\partial_N E_M{}^A) + (\partial_M \xi^N) E_N{}^A \right] . \tag{1.25b}$$

We now have the following result which may be proven by explicitly writing down all expressions.

<u>Lemma</u>:

(1) $L_\xi E^A = 0$ for ξ = susy generating vector field (1.5) and E^A = local frame (1.23) (i.e. E^A is invariant under susy transformations).

(2) Conversely, $L_\xi E^A = 0$ for the local frame (1.23) implies that ξ is the susy generating vector field (1.5).

Hence the (infinitesimal) susy transformations and translations, $z'^M = z^M + \xi^M$ are characterized by vector fields $\xi = \xi^M \partial_M$ satisfying $L_\xi E^A = 0$ with E^A being the supervielbein (1.23).

Since L_ξ is a derivation on the algebra of differential forms[36)] we have the following corollary of (1):

<u>Corollary</u>:

Under infinitesimal susy transformations (and translations) a p-superform transforms as

$$L_\xi \left(\frac{1}{p!} E^{A_1} \ldots E^{A_p} Q_{A_p \ldots A_1} \right) = \frac{1}{p!} E^{A_1} \ldots E^{A_p} \left(L_\xi Q_{A_p \ldots A_1} \right) \tag{1.26}$$

where

$$L_\xi Q_{A_p \ldots A_1} = i_\xi d Q_{A_p \ldots A_1} = \xi^A D_A Q_{A_p \ldots A_1} .$$

In particular the last equation describes infinitesimal susy transformations of superfields Φ (see section I.11 (ii)),

$$L_\xi \Phi = \xi^A D_A \Phi . \tag{1.27}$$

Furthermore an explicit evaluation of the <u>Lie bracket</u>

$$[L_{\xi_1}, L_{\xi_2}]\,\Phi \equiv [\xi_1, \xi_2]\,\Phi$$
$$= \left[\xi_1^M(\partial_M \xi_2^N) - \xi_2^M(\partial_M \xi_1^N)\right]\partial_N \Phi$$

for two supersymmetry generating vector fields ξ_1, ξ_2 leads to

$$[\xi_1, \xi_2]\,\Phi = 2i\,(\zeta_1\sigma^m\bar{\zeta}_2 - \zeta_2\sigma^m\bar{\zeta}_1)\,\partial_m \tag{1.28}$$

which result is in agreement with the susy algebra relation

$$[\zeta_1 Q, \bar{\zeta}_2\bar{Q}] = 2(\zeta_1\sigma^m\bar{\zeta}_2)\,P_m \quad .$$

The geometric formalism introduced in this section is a very useful device and it will be extensively used in our study of the BRS-algebra of SYM-theories (ch.IV.3).

(iii) Differential calculus with an anholonomic basis of 1-forms

If a p-superform is expressed in terms of the natural basis (dz^M) of $T_z M$ (as in equation (1.19)), its components $Q_{A_p\ldots A_1}$ are mixed by a susy transformation since the 1-forms dz^M are not invariant under these transformations. According to the previous corollary this is not the case for the basis (E^A) and therefore the latter is particularly well suited for the formulation of susy invariant theories. For instance in SYM-theories the spinorial components $\mathbb{F}_{\underline{\alpha\beta}}$ of the curvature 2-form $\mathbb{F} = \frac{1}{2} E^A E^B \mathbb{F}_{BA}$ are constrained to vanish : this condition is not modified by a susy transformation since

$$L_\xi\,(E^A E^B \mathbb{F}_{BA}) = E^A E^B (L_\xi \mathbb{F}_{BA}) \quad .$$

As for the basis (D_A) there is a price to pay for using the invariant basis (E^A). In fact the equation

$$[D_C, D_B] = \mathcal{C}_{CB}{}^A\, D_A \tag{1.29}$$

is easily seen to be equivalent to the following formula for the dual basis E^A:

$$dE^A + \frac{1}{2} E^B E^C \mathcal{C}_{CB}{}^A = 0 \quad . \tag{1.30}$$

This equation expresses the anholonomy of the local frame E^A : while $d(dz^M) = 0$ we have $dE^a \neq 0$ for the invariant frame (1.23) and the corresponding anholonomy coefficients $\mathcal{C}_{CB}{}^{A}$ (given by (1.16)) show up whenever the exterior differential of a p-form $(p \geq 1)$ is taken, e.g. in the components of the YM-curvature 2-form $\mathbb{F} \equiv d\mathbb{A} + \mathbb{A}\mathbb{A}$:

$$\mathbb{F} = \frac{1}{2} E^A E^B \left(D_B \mathbb{A}_A - (-)^{ab} D_A \mathbb{A}_B - [\mathbb{A}_A, \mathbb{A}_B] - \mathcal{C}_{BA}{}^{C} \mathbb{A}_C \right) \tag{1.31}$$

where

$$[\mathbb{A}_A, \mathbb{A}_B] = \mathbb{A}_A \mathbb{A}_B - (-)^{ab} \mathbb{A}_B \mathbb{A}_A \quad .$$

I.6 Canonical structures in the local formulation

The anholonomy coefficients $\mathcal{C}_{CB}{}^{A}$ characterizing the invariant frames (D_A), (E^A) are exactly the structure constants of the susy algebra $\mathfrak{m}$ with respect to the usual basis $(t_A) = (P_a, Q_\alpha, \bar{Q}^{\dot{\alpha}})$:

$$[t_B, t_C] = -\mathcal{C}_{BC}{}^{A} t_A \quad , \quad \mathcal{C}_{BC}{}^{A} = -(-)^{bc} \mathcal{C}_{CB}{}^{A} \tag{1.32}$$

(see (1.1) and (1.16)). Therefore equation (1.30) can be considered as a local (superspace) version of the Maurer-Cartan equation of the group SP. In fact[85] the Maurer-Cartan (or canonical) form ω of SP is a 1-form on SP with values in Lie SP = Lie L $\oplus$ $\mathfrak{m}$ and satisfies the equation

$$d\omega + \frac{1}{2} [\omega, \omega] = 0 \quad . \tag{1.33}$$

If $(t_I) = (t_A, M_{ab})$ denotes the usual basis of Lie SP, then

$$\omega = \omega^I t_I = \omega^A t_A + \omega^{ab} M_{ab}$$

and (1.33) reads

$$\begin{aligned} 0 &= (d\omega^A) t_A + \frac{1}{2} (-)^{bc} \omega^B \omega^C [t_B, t_C] \quad \text{(+ terms depending on } \omega^{ab}) \\ &= \left(d\omega^A + \frac{1}{2} \omega^B \omega^C \mathcal{C}_{CB}{}^{A} \right) t_A \quad \text{(+ terms depending on } \omega^{ab}) \end{aligned} \tag{1.34}$$

where we used (1.32), i.e. the fact that $\mathfrak{m}$ represents a subalgebra and not just a vector subspace of Lie SP. In the following chapters we will describe more precisely the correspondence between the 1-forms ω^A on SP and E^A on M = SP/L and in particular between the equations (1.30) and (1.34).

As a further comment (to which we will also come back in the following) we note that the invariance of the form ω under left translations of the group SP is reflected on the quotient space $M = SP/L$ by the invariance of E^A under susy transformations (eq. (1.24)).

I.7 Majorana spinor notation and transformation properties

In the next two chapters we will use Majorana 4-spinors instead of Weyl 2-spinors in order to simplify the notation. Thus in the following we will rewrite some of the previously obtained relations in terms of these spinors and we will briefly discuss their transformation properties.

In the Weyl basis

$$\gamma^m = \begin{bmatrix} 0 & \sigma^m \\ \bar{\sigma}^m & 0 \end{bmatrix} \tag{1.35}$$

of the Dirac algebra a Majorana spinor contains two Weyl spinors:

$$\xi = \begin{pmatrix} \xi_\alpha \\ \bar{\xi}^{\dot\alpha} \end{pmatrix} \quad , \quad \Theta = \begin{pmatrix} \theta_\alpha \\ \bar{\theta}^{\dot\alpha} \end{pmatrix} \quad , \quad \Psi(x) = \begin{pmatrix} \psi_\alpha(x) \\ \bar{\psi}^{\dot\alpha}(x) \end{pmatrix} . \tag{1.36}$$

In terms of four-spinors the SP algebra reads:

$$\begin{aligned}
&[M_{mn}, M_{lp}] = i(\eta_{nl} M_{mp} - \eta_{np} M_{ml} - \eta_{ml} M_{np} + \eta_{mp} M_{nl}) \\
&[P_m, M_{nl}] = i(\eta_{mn} P_l - \eta_{ml} P_n) \\
&[Q_\alpha, M_{mn}] = \frac{1}{2} (\gamma_{mn})_\alpha{}^\beta Q_\beta \\
&\{Q_\alpha, \bar{Q}_\beta\} = -2 (\gamma^m)_{\alpha\beta} P_m \\
&\{Q_\alpha, Q_\beta\} = [P_m, Q_\alpha] = [P_m, P_n] = 0
\end{aligned} \tag{1.37}$$

where $\bar{Q} = Q^\dagger \gamma^0$ and

$$\gamma_{mn} = \frac{1}{4} [\gamma_m, \gamma_n] = \begin{bmatrix} \sigma_{mn} & 0 \\ 0 & \bar{\sigma}_{mn} \end{bmatrix} .$$

The combined rigid susy and Lorentz transformations (1.4), (1.9) take the form

$$x'^m = x^m + a^m + i\left(\bar{\xi}\gamma^m\Theta\right) + \left(e^{-\lambda}\right)^m{}_n\, x^n$$
$$\Theta' = \Theta + \xi + \left[\exp\left(\frac{i}{4}\lambda^{mn}\gamma_{mn}\right)\right]\Theta \quad . \tag{1.38}$$

and for the left-invariant frames (E^A), (D_A) we have

$$E^a = dx^a - i\,\bar{\Theta}\gamma^a(d\Theta)$$
$$E_\alpha = -d\Theta_\alpha \tag{1.39}$$

resp.

$$D_a = \frac{\partial}{\partial x^a}$$
$$D_\alpha = C_{\alpha\beta}\frac{\partial}{\partial\Theta_\beta} + i(\gamma^m\Theta)_\alpha\partial_m \equiv -\frac{\partial}{\partial\bar{\Theta}_\alpha} + i(\gamma^m\Theta)_\alpha\partial_m \; . \tag{1.40}$$

The 4-spinors (1.36) transform under the reducible bi-spinor representation

$$V(A) = \begin{bmatrix} A & \\ & A^{*-1} \end{bmatrix} \tag{1.41}$$

of the group $SL(2,\mathbb{C})$. This representation and the four-vector representation of $SL(2,\mathbb{C})$ (i.e. the "universal covering homomorphism of the Lorentz group $L_+^\uparrow$"),

$$\Lambda : SL(2,\mathbb{C}) \longrightarrow L_+^\uparrow$$
$$A \longmapsto \Lambda(A)$$

are related by the equation[22)]

$$V(A)^{-1}\gamma^m V(A) = \Lambda(A)^m{}_n\,\gamma^n \tag{1.42}$$

For any $\Lambda \in L_+^\uparrow$ there is an element $A \in SL(2,\mathbb{C})$ such that $\Lambda(A) = \Lambda$; thereby[*)] the representation $V(A)$ of $SL(2,\mathbb{C})$ gives rise to a representation $S(\Lambda) := V(A)$ of $L_+^\uparrow$,

---

[*)] For further details see e.g. ref. 157).

$$\Psi'(x') = S(\Lambda)\Psi(x)$$

and from equation (1.42) we have the general relation

$$S(\Lambda)^{-1}\gamma^m S(\Lambda) = \Lambda^m{}_n \gamma^n \tag{1.43}$$

which will be repeatedly used below. (Of course (1.43) may also be deduced directly from the explicit expression

$$S(\Lambda) = \exp\left(\frac{i}{4}\lambda^{mn}\gamma_{mn}\right) \tag{1.44}$$

following from (1.9).)

The infinitesimal version of equation (1.42) is obtained by writing

$$A_\alpha{}^\beta = [\exp l]_\alpha{}^\beta \quad , \quad \Lambda(A)_a{}^b = [\exp L(l)]_a{}^b$$

and expanding (1.42) to first order in ℓ while using the representation (1.35):

$$\sigma^a_{\alpha\dot\alpha}\,\sigma^b_{\beta\dot\beta}\,L_{ab} = -2\,\varepsilon_{\alpha\beta}\,l_{\dot\alpha\dot\beta} + 2\,\varepsilon_{\dot\alpha\dot\beta}\,l_{\alpha\beta} \,. \tag{1.45}$$

Here $L_{ab} = -L_{ba}$ and $\ell_{\alpha\beta} = \ell_{\beta\alpha}$, $\ell_{\dot\alpha\dot\beta} = \ell_{\dot\beta\dot\alpha}$ (see (1.9) and recall $(\sigma_{ab})_\alpha{}^\gamma\,\varepsilon_{\gamma\beta} = (\sigma_{ab})_\beta{}^\gamma\,\varepsilon_{\gamma\alpha}$). Equation (1.45) represents the familiar relation between the generators of the Lorentz group and those of $SL(2,\mathbb{C})$.

I.8 Matrix representation of the SP-group and Maurer-Cartan forms

Following T. Regge[115)] we will give a linear representation of the SP-group in terms of 9×9 matrices and use it to determine the Maurer-Cartan forms of SP[*)]. As we will see in the next chapter the latter can be used to define the canonical linear connection on rigid superspace. Before considering the nine dimensional representation of the SP-group we briefly recall the linear five dimensional representation of the ordinary Poincaré group.

(i) Matrix representation of the Poincaré group

The 4-dimensional Poincaré group $(L,\cdot)\otimes(\mathbb{R}^4,+)$ can be realized linearly by 5×5 matrices

---

*) We note that these forms can also be determined explicitly without using such a representation (see ref. 102).

$$(\Lambda, a) = \begin{bmatrix} \Lambda & a \\ 0 & 1 \end{bmatrix} \begin{matrix} 4 \\ 1 \end{matrix} \quad .$$

(columns of widths 4 and 1)

We can think of these matrices as acting on the real 4-dimensional vector space $\{\binom{x}{1}, x \in \mathbb{R}^4\}$ whose linear structure is defined by

$$\alpha \begin{pmatrix} x \\ 1 \end{pmatrix} + \beta \begin{pmatrix} y \\ 1 \end{pmatrix} = \begin{pmatrix} \alpha x + \beta y \\ 1 \end{pmatrix} \qquad (x, y \in \mathbb{R}^4 \ ; \ \alpha, \beta \in \mathbb{R}).$$

This action,

$$\begin{bmatrix} \Lambda & a \\ 0 & 1 \end{bmatrix} \begin{pmatrix} x \\ 1 \end{pmatrix} = \begin{pmatrix} \Lambda x + a \\ 1 \end{pmatrix} .$$

then describes an <u>affine motion of $\mathbb{R}^4$</u> and the matrix multiplication

$$(\Lambda_1, a_1) \cdot (\Lambda_2, a_2) = (\Lambda_1 \circ \Lambda_2 \ , \ \Lambda_1 a_2 + a_1) \tag{1.46}$$

corresponds to the familiar composition of affine motions. The inverse of the group element (Λ, a) is given by $(\Lambda^{-1}, -\Lambda^{-1}a)$.

(ii) <u>Matrix representation of the SP-group</u>

The 4-dimensional SP-group can be realized by 9×9 matrices

$$(\Lambda, a, \xi) = \begin{bmatrix} \Lambda & i\left(\bar{\xi} \gamma S(\Lambda)\right) & a \\ 0 & S(\Lambda) & \xi \\ 0 & 0 & 1 \end{bmatrix} . \tag{1.47}$$

Here $\Lambda \in L$, $a \in \mathbb{R}^4$ and ξ is a global Majorana spinor ; the 4×4 submatrices of (1.47) have the following index structure:

$$\Lambda^a{}_b \quad , \quad S(\Lambda)^\alpha{}_\beta \quad , \quad i\left(\bar{\xi} \gamma^a S(\Lambda)\right)_\beta \quad .$$

Upon use of equation (1.43) and of

$$S(\Lambda)\, S(M) = S(\Lambda M)$$

$$\gamma^0 S(\Lambda)^{-1} \gamma^0 = S(\Lambda)^\dagger \quad , \qquad (\gamma^0)^2 = \mathbb{1}_4 \quad ,$$

the usual matrix multiplication reproduces the SP-group relations

$$(\Lambda, x, \xi)(M, y, \eta) = \left(\Lambda M,\ x + \Lambda y + i\bar{\xi}\gamma S(\Lambda)\eta,\ \xi + S(\Lambda)\eta\right) \quad (1.48)$$

The unit element is represented by $(\mathbb{1}_4, 0, 0)$ and the inverse of (Λ, x, ξ) is given by

$$(\Lambda, x, \xi)^{-1} = \left(\Lambda^{-1},\ -\Lambda^{-1}x + i\Lambda^{-1}(\bar{\xi}\gamma\xi),\ -S(\Lambda)^{-1}\xi\right). \quad (1.49)$$

Analogously to the ordinary Poincaré group the matrix (Λ, a, ξ) can be thought of as acting on a (4,4)-dimensional supervector space*):

$$\begin{bmatrix} \Lambda & i\bar{\xi}\gamma S(\Lambda) & a \\ 0 & S(\Lambda) & \xi \\ 0 & 0 & 1 \end{bmatrix}\begin{bmatrix} x \\ \Theta \\ 1 \end{bmatrix} = \begin{bmatrix} \Lambda x + i\bar{\xi}\gamma S(\Lambda)\Theta + a \\ S(\Lambda)\Theta + \xi \\ 1 \end{bmatrix}.$$

In particular for $\Lambda = \mathbb{1}_4$ this action describes the rigid susy transformations (and ordinary translations):

$$(\mathbb{1}, a, \xi)\begin{bmatrix} x \\ \Theta \\ 1 \end{bmatrix} = \begin{bmatrix} x + a + i\bar{\xi}\gamma\Theta \\ \Theta + \xi \\ 1 \end{bmatrix}.$$

(iii) Maurer-Cartan formson SP and their pullbacks to superspace

We recall that the Maurer-Cartan form ω of a Lie group G is a 1-form on the manifold G with values in the associated Lie algebra $\mathfrak{g}$: $\omega \in \Lambda^1(G, \mathfrak{g})$. For matrix Lie groups it is given by $\omega = g^{-1}dg$[73]; it is invariant under left multiplication in the group,

$$l_a^* \omega = (a\cdot g)^{-1} d(a\cdot g) = g^{-1}(a a^{-1})\, dg = \omega \qquad (a \in G)$$

and satisfies the Maurer-Cartan equation (1.33) which, for matrix groups, can be written as

---

*) For a formal definition see e.g. ref. 42).

$$d\omega + \omega \wedge \omega = 0 \quad . \tag{1.50}$$

In particular for the SP-group we have:

$$\omega = (\Lambda, x, \Theta)^{-1} d(\Lambda, x, \Theta) = \begin{bmatrix} \Lambda^{-1} d\Lambda & i\Lambda^{-1}(d\bar{\Theta}\gamma S) & \Lambda^{-1}(dx - i\bar{\Theta}\gamma d\Theta) \\ 0 & S^{-1} dS & S^{-1}(d\Theta) \\ 0 & 0 & 0 \end{bmatrix} \tag{1.51}$$

Here all submatrices can be expressed in terms of the following 1-forms:

$$\begin{aligned} \Omega^a{}_b &= (\Lambda^{-1} d\Lambda)^a{}_b \\ V^a &= (\Lambda^{-1}(dx - i\bar{\Theta}\gamma d\Theta))^a \\ \Psi_\alpha &= (S(\Lambda)^{-1} d\Theta)_\alpha \quad . \end{aligned} \tag{1.52}$$

Indeed from the relation (1.43) it follows that

$$i\Lambda^{-1}(d\bar{\Theta}\gamma S) = i\bar{\Psi}\gamma \quad ,$$

while Schur's lemma applied to the Dirac algebra leads to the general formula

$$S^{-1} dS = \frac{1}{2}\Omega_{ab}\gamma^{ab}$$

where

$$\Omega_{ab} := \eta_{ac}\Omega^c{}_b \qquad (= -\Omega_{ba}) \quad .$$

Thus the Maurer-Cartan equation for (1.51) corresponds to the following set of equations for Ω, V and ψ (we suppress the symbol $\wedge$ for the exterior multiplication):

$$\begin{aligned} d\Omega + \Omega\Omega &= 0 \\ dV + \Omega V + i\bar{\Psi}\gamma\Psi &= 0 \\ d\Psi + \frac{1}{2}(\Omega_{ab}\gamma^{ab})\Psi &= 0 \quad . \end{aligned} \tag{1.53}$$

Note that these relations reflect the semi-direct sum structure (1.2) of the SP-algebra.

The left invariance of the form (1.51),

$$l^{*}_{(M,a,\xi)}\,\omega = \omega$$

is equivalent to the left-invariance of the forms Ω, V and ψ :

$$l^{*}_{(M,a,\xi)} \begin{Bmatrix} \Omega \\ V \\ \psi \end{Bmatrix} = \begin{Bmatrix} \Omega \\ V \\ \psi \end{Bmatrix} . \tag{1.54}$$

These forms which are defined on the "SP-group manifold" can be pulled back to the quotient space by a <u>(local) section</u> $\sigma : M \rightarrow SP$, i.e. by the map

$$\begin{aligned} \sigma^{*} : \Lambda^{1}(SP, \mathcal{L}ie\, SP) &\longrightarrow \Lambda^{1}(M, \mathcal{L}ie\, SP) \\ \omega &\longmapsto \sigma^{*}\omega \end{aligned} \tag{1.55}$$

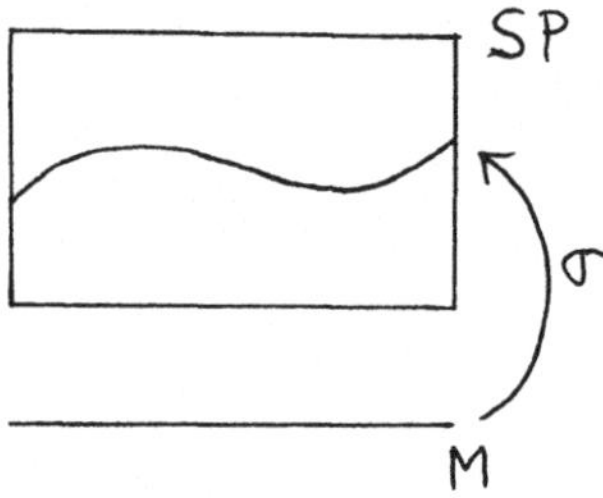

For a constant section σ this operation amounts to setting $\Lambda = \mathbb{1}_4$, $d\Lambda = 0$ so that one obtains the following equations from (1.52):

$$\begin{aligned} \sigma^{*}\Omega &= 0 \\ (\sigma^{*}V)^{a} &= dx^{a} - i\,\bar{\Theta}\gamma^{a}d\Theta = E^{a} \\ (\sigma^{*}\psi)_{\alpha} &= -\,d\Theta_{\alpha} = E_{\alpha} \quad . \end{aligned} \tag{1.56}$$

Here we recognize the left-invariant frame (1.39).

The invariance of the Maurer-Cartan 1-form ω under left translations of the SP-group (eq. (1.54)) is reflected on superspace by the invariance of the vielbein forms $E^{a} = \sigma^{*}V^{a}$, $E_{\alpha} = \sigma^{*}\psi_{\alpha}$ under translations and susy transformations, i.e. the transformations induced on SP/L by the left action of the group elements $G(a,\xi\,;0)$ (eq. (1.24)).

Since the pull-back map (1.55) is an algebra-morphism and commutes with the exterior differentiation,

$$\sigma^*(\alpha \wedge \beta) = (\sigma^*\alpha) \wedge (\sigma^*\beta) \ , \qquad \sigma^* d\alpha = d\sigma^*\alpha \ ,$$

the Maurer-Cartan equations (1.53) can be directly pulled back to rigid superspace. In terms of Weyl 2-spinors the resulting equations read

$$dE^a - 2i\, \sigma^a_{\alpha\dot\beta}\, E^\alpha E^{\dot\beta} = 0$$

$$dE^\alpha = 0 = dE_{\dot\alpha} \ .$$

These are the previously obtained equations (1.30) which describe the anholonomy of the frame (E^A). Like the Maurer-Cartan equations (1.53) they reflect the semi-direct sum structure of the SP-algebra: the coefficients $2\sigma^a_{\alpha\dot\beta}$ are the structure constants of the susy algebra $\mathcal{M}$.

(iv) <u>Left-invariant vector fields on SP and their projections to superspace</u>

When labeling the elements $(\Lambda^a{}_b)$ of the Lorentz group one has to pay attention to the position of the different indices. In order to avoid this technical complication in the following we will consider the orthogonal group $O(4)$ instead of the Lorentz group $L = O(3,1)$; thus we have

$$(\Lambda^{-1})^{ab} = (\Lambda^t)^{ab} = \Lambda^{ba} \qquad (\Lambda \in O(4)) \ .$$

At any point (Λ, x, Θ) of the "SP group manifold" the 1-forms Ω^{ab}, V^a and ψ_α span a basis of the cotangent space for which the dual is given by[102)]

$$\begin{aligned} \mathbb{D}_{ab} &= (\Lambda^{-1})^{ac}\frac{\partial}{\partial\Lambda^{cb}} - (\Lambda^{-1})^{bc}\frac{\partial}{\partial\Lambda^{ca}} = \Lambda^{ca}\frac{\partial}{\partial\Lambda^{cb}} - \Lambda^{cb}\frac{\partial}{\partial\Lambda^{ca}} \\ \mathbb{D}_a &= (\Lambda^{-1})^{ac}\frac{\partial}{\partial x^c} = \Lambda^{ca}\frac{\partial}{\partial x^c} \\ \mathbb{D}_\alpha &= (S(\Lambda)^{-1})_{\alpha\beta}\left[C_{\beta\gamma}\frac{\partial}{\partial\Theta_\gamma} + i\,(\gamma^m_{\beta\gamma}\Theta_\gamma)\,\partial_m \right] . \end{aligned} \tag{1.57}$$

Indeed we have:

$$\Omega^{ab}(\mathbb{D}_c) = 0 = \mathbb{D}_\alpha(\overleftarrow{\Omega}^{ab})$$

$$\Omega^{ab}(\mathbb{D}_{cd}) = \delta^{ac}\delta^{bd} - \delta^{ad}\delta^{bc}$$

$$V^a(\mathbb{D}_{bc}) = 0 = \mathbb{D}_\alpha(\overleftarrow{V}^a)$$

$$V^a(\mathbb{D}_b) = \delta^{ab}$$

$$\mathbb{D}_a(\overleftarrow{\Psi}_\alpha) = 0 = \mathbb{D}_{ab}(\overleftarrow{\Psi}_\alpha) .$$

Under the push forward π_* induced by the natural projection map $\pi : SP \to SP/L$ the tangent vectors (1.57) of SP yield the familiar left-invariant vector fields (1.40) of rigid superspace:

$$\pi_* \mathbb{D}_{ab} = 0$$

$$\pi_* \mathbb{D}_a = \frac{\partial}{\partial x^a} = D_a$$

$$\pi_* \mathbb{D}_\alpha = C_{\alpha\beta}\frac{\partial}{\partial \Theta_\beta} + i(\gamma^m_{\alpha\beta}\Theta_\beta)\partial_m = D_\alpha .$$

Thus we have shown that there is a very close relationship between the canonically given geometric structures on SP and the familiar ones on SP/L. In the next chapter we will see that these structures can be given a further reaching significance and geometric interpretation.

I.9 Invariant connections on reductive homogeneous spaces

In order to describe the parallel displacement of tangent vectors on a manifold M one has to introduce some linear connection. For reductive homogeneous spaces G/H these connections have been classified by H.C. Wang[148]; in particular there is a canonically given G-invariant one which we will consider below. These results are exposed in great generality in the textbook of Kobayashi and Nomizu[85] and in the first section we will give a simple and short presentation of the specific result we are interested in. Although super fibre bundles have recently been defined[31] in the framework of Rogers-Jadczyk-Pilch-supermanifolds[116),82] we will not attempt a formal generalization of the ordinary space results in view of the remarks made in the introduction. In the second section it will be shown how the familiar local superspace geometry follows from the general theory and how it can be interpreted from this view-point.

(i) The main result

If G is a Lie group with a closed subgroup H, then G can be viewed as a principal fibre bundle with structure group H over the homogeneous space $M = G/H$ (H acts on G by multiplication from the right). Now suppose that, in addition, G is connected and that the quotient space M is reductive, i.e. the Lie algebra $\mathfrak{g}$ of G can be decomposed into a vector space direct sum of the Lie algebra $\mathfrak{h}$ of H and an ad(H)-invariant subspace $\mathfrak{m}$:

$$\mathfrak{g} = \mathfrak{h} + \mathfrak{m} \tag{1.58}$$

$$\mathrm{ad}(H)\,\mathfrak{m} \subset \mathfrak{m} .$$

Then the h-component $\omega^{(h)}$ (w.r.t. the decomposition (1.58)) of the Maurer-Cartan 1-form ω of G defines a connection in the bundle G*). This canonically given connection is invariant by the left translations of G, i.e.

$$l_g^{*}\,\omega^{(h)} = \omega^{(h)} \qquad (g \in G) .$$

Under the identification $\mathfrak{g} \cong T_eG$, with e denoting the identity of G, the subspace $\mathfrak{m}$ of $\mathfrak{g}$ corresponds to the horizontal subspace of the connection $\omega^{(h)}$ at e.

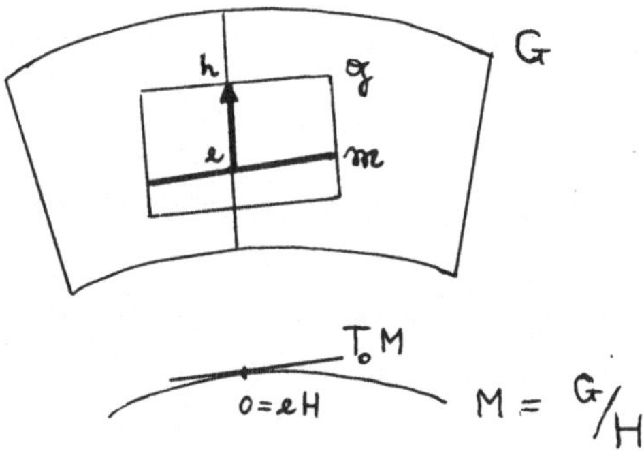

Although we are naturally given a connection in the bundle G over M this is not really what we need. Indeed for describing the parallel transport of tangent vectors on M one needs a linear connection, i.e. a connection in the bundle $L(M)$ of linear frames on M. The latter is a principal fibre bundle over M with structure group $GL(n,\mathbb{R})$ (where $n = \dim M$). The fibre $L(M)_x$ over a point $x \in M$ is the set of all linear frames at x; such a frame u can be viewed either

*) The assumption that G/H is reductive garanties that the h-valued 1-form $\omega^{(h)}$ transforms correctly, i.e. by the adjoint representation of the structure group H.

as an ordered basis $X_1,\ldots,X_n$ of the tangent space T_xM or equivalently as an isomorphism of the vector spaces $\mathbb{R}^n$ and T_xM,

$$u : \mathbb{R}^n \longrightarrow T_x M \quad .$$

(The relation between these two definitions is given by $u(e_i) = X_i$ where $(e_1,\ldots,e_n)$ denotes the natural basis of $\mathbb{R}^n$.)

The left action of G on itself induces in a natural manner an action of G on the quotient space $M = G/H$ and on the bundle $L(M)$; in fact

$$\begin{aligned} l_g : G &\longrightarrow G \\ g_0 &\longmapsto g\cdot g_0 \end{aligned}$$

induces the diffeomorphism

$$\begin{aligned} f_g : M &\longrightarrow M \\ g_0 H &\longmapsto f_g(g_0 H) = (g g_0) H \end{aligned}$$

and the latter yields the automorphism

$$\begin{aligned} \tilde{f}_g : L(M) &\longrightarrow L(M) \\ u = (X_1,\ldots,X_n) &\longmapsto \tilde{f}_g(u) = \left((f_g)_* X_1 , \ldots , (f_g)_* X_n \right) , \end{aligned}$$

i.e. a linear frame u at $x \in M$ is mapped into a linear frame $\tilde{f}_g(u)$ at $f_g(x)$ by using the projection $(f_g)_*$ of tangent vectors.

We now fix an arbitrary linear frame $u_o \in L(M)$ at $0 \equiv eH \in M$ and recall the definition of the <u>linear isotropy representation</u> of $H \subset G$: if we identify T_oM with $\mathbb{R}^n$ by the frame $u_o : \mathbb{R}^n \to T_oM$, the linear isotropy representation of H on the vector space $\mathbb{R}^n$ can be identified with the homomorphism

$$\begin{aligned} \lambda : H &\longrightarrow GL(n,\mathbb{R}) \\ h &\longmapsto \lambda(h) = u_o^{-1} \circ (f_h)_* \circ u_o \end{aligned} \tag{1.59}$$

where $(f_h)_* : T_oM \to T_oM$ denotes the projection of tangent vectors at $0 \in M$.

Wang's classification theorem implies the existence of a canonical connection on $L(M)$ which can be characterized as follows.

Main result:

If we set $P := \{\tilde{f}_g(u_o);\ g \in G\}$, then P is a subbundle of $L(M)$ with structure group $\lambda(H) \subset GL(n,\mathbb{R})$ and is isomorphic to the principal fibre bundle G (with structure group H and base manifold $M = G/H$). Under this isomorphism which we denote by F the group element $g \in G$ is mapped onto $\tilde{f}_g(u_o) \in P$ and the G-invariant connection $\omega^{(h)}$ in G is mapped into a G-invariant connection in P which is the restriction to P of the canonical connection in L(M) (w.r.t. the decomposition $\mathfrak{g} = h + \mathfrak{m}$)*).

Figure:

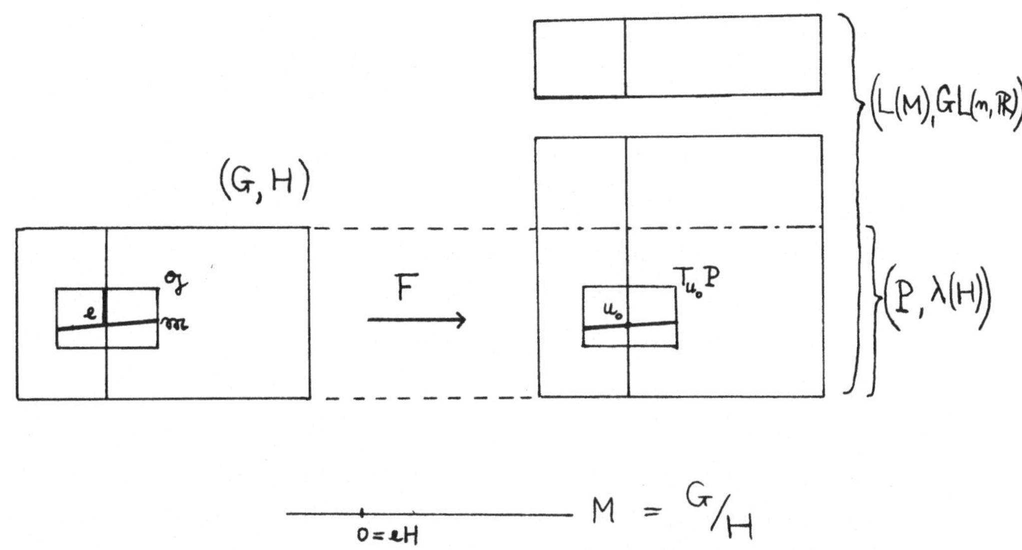

The geometric interpretation of the canonical linear connection is summarized in appendix 4.

We remark that for $G = P_o$, i.e. the identity component of the Poincaré group in n dimensions and $H = L_o$ we can identify the coset space $M = P_o/L_o$ with $\mathbb{R}^n$ and the linear isotropy representation λ with the identity map $id : L_o \to L_o \subset GL(n,\mathbb{R})$. In this case $(P, \lambda(H))$ corresponds to the reduction of the principal fibre bundle $(L(M), GL(n,\mathbb{R}))$ to one of the four components of the bundle $O(M)$ of orthonormal frames (whose structure group is the Lorentz group L)[21]; furthermore in this example the canonical linear connection corresponds to the canonical flat connection in $P = M \times L_o$.

---

*) That one connection is mapped into another means that the corresponding horizontal spaces are mapped into one another (by F_*) ; G-invariance of a connection Ω in P means $(\tilde{f}_g)^*\Omega = \Omega$ for all $g \in G$).

(ii) Local (superspace) expressions

To obtain a local expression on M for the canonical linear connection one has to pull back the corresponding connection 1-form Ω from P to M by a local section $s : U \to L(M)$ (defined on an open subset $U \subset M$). Because of the preceding theorem this is equivalent to pulling back the connection form $\omega^{(h)}$ from G to M by a local section $\sigma : U \to G$. For the discussion of this procedure we specify the notation to the case where $M = G/H$ is the rigid superspace, $M = SP_o/L_o$ *). Thus we have the decomposition

$$\omega = \omega^{(m)} + \omega^{(h)} = \omega^A t_A + \omega^{cd} M_{cd} \qquad (t_A \in \{P_a, Q_\alpha, \bar{Q}^{\dot{\alpha}}\})$$

and consequently

$$\sigma^*\omega = \sigma^*\omega^{(m)} + \sigma^*\omega^{(h)} = (\sigma^*\omega)^A t_A + (\sigma^*\omega)^{cd} M_{cd}$$

$$=: dz^M \left(E_M{}^A(z)\, t_A + \varphi_M{}^{cd}(z)\, M_{cd} \right)$$

where we defined the following 1-forms on M :

$$E^A \equiv dz^M E_M{}^A \equiv (\sigma^*\omega)^A$$
$$\varphi^{cd} \equiv dz^M \varphi_M{}^{cd} \equiv (\sigma^*\omega)^{cd} . \qquad (1.60)$$

The Lorentz algebra valued 1-form $\varphi^{cd} M_{cd}$ represents a <u>local expression for the canonical linear connection</u> and E^A is a local frame which is automatically given by this construction ; as a consequence of the left-invariance of ω both φ^{cd} and E^A are invariant under susy transformations.

Note that the <u>Lorentz connection</u>

$$dz^M \left(\varphi_M{}^{cd} M_{cd}\right)_b{}^a \equiv dz^M (\phi_M)_b{}^a \equiv \phi_b{}^a$$

not only acts on the spatial tangent space indices, but also on the spinorial ones. In fact from the general relation (1.45) between the generators of the Lorentz group and of $SL(2,\mathbb{C})$ we have the connection forms

$$\left(\phi_B{}^A\right) = \left(\phi_b{}^a ,\ \phi_\beta{}^\alpha ,\ \phi^{\dot{\beta}}{}_{\dot{\alpha}}\right) \qquad (1.61)$$

---

*) In the following considerations we will not use the fact that Lie SP_o has a semi-direct sum structure.

with

$$\phi_\beta{}^\alpha = -\frac{1}{2}\phi_{ba}(\sigma^{ba})_\beta{}^\alpha \;, \qquad \phi^{\dot\beta}{}_{\dot\alpha} = -\frac{1}{2}\phi_{ba}(\bar\sigma^{ba})^{\dot\beta}{}_{\dot\alpha} \;.$$

The expressions (1.60),(1.61) take a particularly simple form for a <u>parallel section</u> σ, i.e. a section for which the push forward $\sigma_*: T_zM \to T_{\sigma(z)}(SP_o)$ maps vectors from T_zM into the horizontal subspace of $T_{\sigma(z)}(SP_o)$.

<u>Figure</u>:

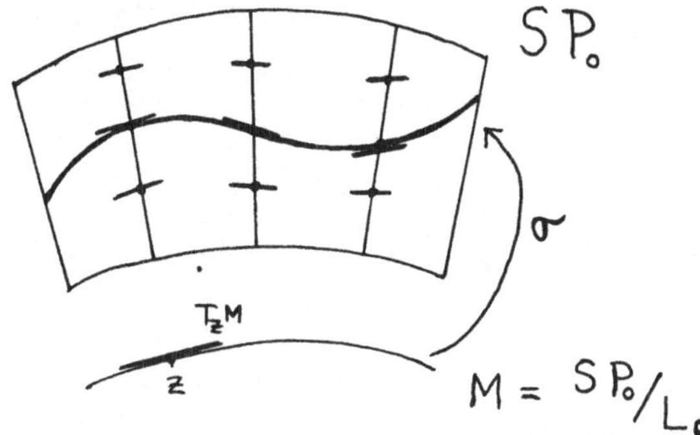

In fact by the very definition of the connection $\omega^{(h)}$ on $G = SP_o$ this implies that the local representative $\sigma^*\omega^{(h)}$ vanishes identically:

$$(\sigma^*\omega^{(h)})(X) \equiv \omega^{(h)}(\sigma_* X) = 0 \quad \text{for all} \quad X \in T_zM \;.$$

Explicit expressions for the corresponding vielbein forms $E^A = (\sigma^*\omega)^A$ have been obtained in section I.8 (iii) (see eqs (1.56)): these represent precisely the familiar invariant basis of T_z^*M.

Of course the differential operators (D_A) which represent its dual can be interpreted as local expressions for the <u>Lorentz covariant derivative</u> w.r.t. the invariant connection $\phi_B{}^A \equiv 0$:

$$\mathcal{D}_A = E_A{}^M(\partial_M + \phi_M) = E_A{}^M \partial_M = D_A \tag{1.62}$$

Whereas the <u>curvature</u>

$$R[\phi]_B{}^A = d\phi_B{}^A + \phi_B{}^C \phi_C{}^A$$

of the connection $\phi \equiv 0$ vanishes identically on superspace, the torsion does not: when expressing the <u>torsion</u> 2-form of ϕ with respect to the local basis E^A,

$$T[\phi]^A = \mathcal{D}E^A = dE^A + E^B\phi_B{}^A \quad ,$$

we obtain for the connection $\phi \equiv 0$:

$$T^A[0] \equiv \frac{1}{2} E^B E^C T_{CB}{}^A[0] = dE^A . \tag{1.63}$$

Comparison of the last equality with the two equivalent equations (1.29),(1.30) leads to the relation

$$T_{CB}{}^A[0] = -[D_C , D_B]^A . \tag{1.64}$$

(This expression corresponds to the usual equation for the commutator of Lorentz covariant derivatives:

$$[\mathcal{D}_C , \mathcal{D}_B] V_D = - R_{CBD}{}^A V_A - T_{CB}{}^A \mathcal{D}_A V_D \tag{1.65}$$).

Thus we see that the components $T_{CB}{}^A[0]$, which are usually referred to as the components of the "flat superspace torsion", actually express the anholonomy of the invariant frame (D_A). We note that equation (1.64) has an intuitive geometric interpretation*) which is characteristic for the canonical linear connection on a reductive homogeneous space (see appendix 4).

As emphasized above the foregoing discussion locally applies to any reductive homogeneous space $M = G/H$ with decomposition $\mathfrak{g} = h + \mathfrak{m}$; in our particular case where $\mathfrak{m}$ is actually a (graded) Lie algebra with structure constants $\mathcal{C}_{CB}{}^A$, the components (1.64) are constant functions on M and coincide with $-\mathcal{C}_{CB}{}^A$:

$$T_{\dot\beta\alpha}{}^c[0] = T_{\alpha\dot\beta}{}^c[0] = 2i\,\sigma^c_{\alpha\dot\beta} = -\{D_\alpha , \bar{D}_{\dot\beta}\}^c \tag{1.66}$$

(other components = 0).

(iii) Is rigid superspace a flat space?

The torsion and the (corresponding) curvature are not intrinsic properties of a manifold, rather they represent mathematical entities associated to a linear connection which has to be chosen for describing the parallel displacement of tangent vectors on this space. Thus mathematicians consider so-called affine manifolds[74], i.e. manifolds supplemented with a specific linear connection**). For these objects it makes sense to talk about a space with torsion and/or curvature

---

*) For the geometric interpretation of the torsion and of the Lie bracket see for instance refs 101) and 73).

**) The terms "linear connection" and "affine connection" are used interchangeably in the literature ; in fact they represent equivalent concepts[85].

or about a flat space : by definition an affine manifold is flat, if the associated curvature and torsion forms (or tensors) vanish identically. According to this definition the set $\mathbb{R}^n$ supplemented with the standard differential structure and the canonical flat connection ($\phi \equiv 0$ on $\mathbb{R}^n$) is a flat space : the parallel transport of vectors by this connection corresponds to the usual Euclidean notion of parallelism. On the other hand rigid superspace supplied with the canonical linear connection defined in the last section is a space with torsion and thus not a flat space.

(iv) Christoffel symbols of the canonical connection

There is a 1-1-correspondence between the set of linear connections on a manifold M and the set of covariant derivatives on the tangent bundle of M (ref. 85, Prop. 2.8 and 7.5): the connection form $\phi^a{}_b$ representing locally on M a linear connection and the Christoffel symbols $(\Gamma_\ell)^m{}_n$ describing a covariant derivative with respect to a local coordinate system $\{x^m\}$ are related through a formal gauge transformation given by some vielbein $e^a{}_m$ (ref. 45, p.246):

$$\begin{aligned} \Gamma^m{}_n &= \left(e^{-1}\phi e + e^{-1} de\right)^m{}_n \\ &= e^m{}_a \phi^a{}_b e^b{}_m + e^m{}_a de^a{}_m . \end{aligned} \tag{1.67}$$

The appropriate superspace generalization of this formula reads : ($\Gamma_N{}^M \equiv dz^L \Gamma_{LN}{}^M$, $\phi_A{}^B \equiv dz^L(\phi_L)_A{}^B$)

$$\Gamma_{LN}{}^M = (-)^{\ell(a+n)}\left\{E_N{}^A (\phi_L)_A{}^B E_B{}^N - E_N{}^A (\partial_L E_A{}^M)\right\}.$$

Thus the Christoffel symbols associated to the canonical linear connection $\phi \equiv 0$ and to the invariant basis (E^A) are given by*)

$$\Gamma_{\mu\nu}{}^m = -i\,(\gamma^m)_{\nu\mu} \qquad \text{(other components = 0)}\cdot \tag{1.68}$$

These quantities represent one member of the family of invariant affine connections determined by B. de Witt who used a very different approach (ref. 42, eq.(4.4.39)).

These remarks essentially close our account of the geometry of rigid superspace. Before adding a few related comments we will compare the previously described standard approach with the Riemannian one which might be more familiar from the study of other physical theories.

---

*) We use the Majorana spinor notation (1.39).

I.10 Rigid superspace from the Riemannian view-point

Non-supersymmetric field theories are usually defined on some (pseudo)-Riemannian space (M,g), i.e. on a manifold M which is equipped with some (pseudo-) Riemannian metric g. Thus it seems natural to attempt a generalization of this framework to supersymmetric theories and to introduce some supermetric on rigid superspace SP/L in order to "extend" the Minkowski metric of the underlying body manifold $\mathbb{R}^4$. Such a metric has actually been proposed shortly after the invention of superspace[159),129)]; thereafter it was considered in particular by R. Arnowitt and P. Nath as a vacuum solution in their Riemannian approach to supergravity[*)]. Although this point of view might eventually be acceptable for the study of globally supersymmetric theories it turns out that rigid superspace does not fit naturally as a special case of Riemannian supergravity[162)]: a certain singular limit has to be taken for making contact between Riemannian supergravity and the usual vielbein (i.e. affine space) approach to supergravity[10),11)]. Nevertheless it is both interesting and instructive to compare this approach and the corresponding results with the familiar ones.

(i) Definition of a supermetric[159),129),42)]

It seems natural to generalize the Minkowski metric η by the following supermetric:

$$(\eta_{AB}) = \begin{bmatrix} \eta_{ab} & & \\ & K\varepsilon_{\alpha\beta} & \\ & & K\,\varepsilon^{\dot{\alpha}\dot{\beta}} \end{bmatrix} . \qquad (1.69)$$

In fact this covariant second rank tensor is graded symmetric and allows to lower anholonomic superindices. (The raising of superindices can be done with the inverse matrix). The constant K occuring in (1.69) has to be introduced for dimensional reasons as will be seen below.

(η_{AB}) is to be interpreted as a super tangent space metric. If $E^A = dz^M E_M{}^A$ denotes the left-invariant vielbein forms, the rigid superspace metric g may be defined by the line element

$$\begin{aligned} ds^2 = E^A \eta_{AB} E^B &= \left(dx^a - i\,(d\theta)\sigma^a\bar{\theta} + i\,\theta\sigma^a d\bar{\theta}\right)^2 + K\,(d\theta)(d\theta) + K\,(d\bar{\theta})(d\bar{\theta}) \\ &=: dz^M g_{MN}\, dz^N \end{aligned} \qquad (1.70)$$

---

*) A review of this work is given in ref. 11).

or equivalently by

$$g_{MN} = (-)^{n+nb} E_M{}^A \eta_{AB} E_N{}^B .$$

Alternatively g can be defined as the supermetric which is invariant under translations, supersymmetry and Lorentz transformations, see section (iii). It has the explicit form

$$(g_{MN}) = \begin{bmatrix} g_{mn} = \eta_{mn} & g_{m\nu} = i(\sigma_m\bar{\theta})_\nu & g_m{}^{\dot{\nu}} = i(\bar{\sigma}_m\theta)^{\dot{\nu}} \\ g_{\mu n} = -i(\sigma_n\bar{\theta})_\mu & g_{\mu\nu} = \varepsilon_{\mu\nu}[K+2(\bar{\theta}\bar{\theta})] & g_\mu{}^{\dot{\nu}} = 2\theta_\mu\bar{\theta}^{\dot{\nu}} \\ g^{\dot{\mu}}{}_n = -i(\bar{\sigma}_n\theta)^{\dot{\mu}} & g^{\dot{\mu}}{}_\nu = 2\bar{\theta}^{\dot{\mu}}\theta_\nu & g^{\dot{\mu}\dot{\nu}} = \varepsilon^{\dot{\mu}\dot{\nu}}[K+2(\theta\theta)] \end{bmatrix} \qquad (1.71)$$

and the following symmetry properties:

$$g_{NM} = (-)^{n+m+nm} g_{MN} .$$

We remark that there are two distinct, although related types of covariant supertensors (lower superindices) ; for a detailed discussion of this point and the definition of the corresponding inverse metrics (g^{MN}) we refer to an article of S.S. Chang[33].

In natural units

$$[dx^a] = -1 \quad , \quad [d\theta^\alpha] = -\tfrac{1}{2} = [d\bar{\theta}_{\dot{\alpha}}] \qquad (1.72)$$

and therefore eq.(1.70) implies $[K] = -1$.

The constant K has the following interpretation. If the determinant of (g_{MN}) is defined in terms of the supertrace[8),42)],

$$\det(g_{MN}) = \exp[\operatorname{str} \ln(g_{MN})] ,$$

the <u>metric density</u> $\sqrt{-g} \equiv \sqrt{-\det(g_{MN})}$ is given by

$$\sqrt{-g} = \frac{1}{K^2} . \qquad (1.73)$$

In the Riemannian approach to supergravity the metric (1.71) occurs as a vacuum state of gauged supersymmetry, $\langle 0|g_{MN}|0\rangle$; in this framework the singular limit $K \to 0$ has to be considered if one wants to make contact with the usual vielbein approach to supergravity[10),11)].

(ii) The Riemannian connection

On (pseudo-)Riemannian manifolds (M,g) there is a distinguished linear connection, namely the Levi-Civita or Riemannian one : it is the unique linear connection which is torsionless (symmetric w.r.t. holonomic frames) and metric (i.e. the parallel transport by this connection along a smooth path on M preserves the scalar product g). For rigid superspace defined as a super-Riemannian space this connection has been determined and its curvature has been evaluated[159),129),42)]: it does not vanish, rather it depends on the fermionic variables θ and $\bar{\theta}$. Thus our Riemannian superspace does not represent a flat space, but a curved "extension" of ordinary Minkowski space $(\mathbb{R}^4,\eta)$[*)].

(iii) Relation between Killing vector fields and susy transformations

Before discussing the Killing vector fields associated to the previously defined supermetric, we recall a few general facts about these fields and consider the Minkowski metric as a simple example.

On a (pseudo-) Riemannian manifold (M,g) there is a particularly interesting class of vector fields namely those which generate diffeomorphisms of M which leave the metric g invariant. They are referred to as the Killing vector fields associated to g and they can be characterized by the condition

$$0 = (L_\xi g) = (L_\xi g)_{mn}\, dx^m \otimes dx^n \tag{1.74}$$

where

$$\xi = \xi^m \partial_m$$

and

$$(L_\xi g)_{mn} = \xi^l (\partial_l g_{mn}) + g_{ml}(\partial_n \xi^l) + g_{nl}(\partial_m \xi^l) .$$

By the introduction of local vielbein forms $e^a = e^a{}_m dx^m$ the metric g can be written as $g_{mn} = e^a{}_m e^b{}_n \eta_{ab}$ where $\eta_{ab} = \pm\delta_{ab}$ represents the flat tangent space metric; thus[143)] $\xi = \xi^m \partial_m$ is a Killing vector field with respect to the metric $g = g_{mn}\, dx^m \otimes dx^n = \eta_{ab}\, e^a \otimes e^b$ if and only if

---

*) We recall[36)] that a (pseudo-) Riemannian manifold is said to be flat, if it is isometric to the (pseudo-) Riemannian manifold $\mathbb{R}^n$ with the usual differential structure and metric, i.e. the metric

$$ds^2 = \eta_{mn}\, dx^m\, dx^n$$

where $\eta_{mn} = \delta_{mn}$ (resp. $\pm\delta_{mn}$ in the pseudo-Riemannian case).

$$L_\xi e^a = M^a{}_b \, e^b \tag{1.75}$$

with $M^a{}_b$ satisfying

$$\eta_{ac} M^c{}_b + \eta_{bc} M^c{}_a = 0 \quad .$$

In other words : ξ is a Killing vector field, if its action on the local frame e^a corresponds to an infinitesimal rotation. (In the pseudo-Riemannian case the rotation is either a true or a hyperbolic rotation.)

In the simple case where (M,g) is the ordinary Minkowski space $(\mathbb{R}^4,\eta)$ we have:

$$\eta_{mn} = \delta^a{}_m \, \delta^b{}_n \, \eta_{ab}$$

$$e^a = \delta^a{}_m \, dx^m$$

and the Killing vector fields $\xi = \xi^m \partial_m$ are those which generate the Poincaré transformations:

$$\xi^m(x) = \lambda^{mn} x_n + a^m \qquad (\lambda^{mn} = -\lambda^{nm}) .$$

If we impose on ξ the condition

$$L_\xi e^a = 0$$

which is obviously stronger than the Killing condition (1.75) we select out the vector field $\xi = a^m \partial_m$ which generates the translations.

A similar thing happens in the case of rigid superspace, but due to the fact that the left-invariant vielbein forms E^A depend explicitly on the fermionic coordinates $\theta^\mu, \bar{\theta}_{\dot{\mu}}$ the condition $L_\xi E^A = 0$ does not only select out the ordinary translations, but also the rigid susy transformations from the whole set of infinitesimal transformations which leave the superspace metric (1.71) invariant. The Killing vector fields associated to (1.71) have been determined independently by L. Kannenberg[83] and R. Arnowitt and P. Nath[9] : in terms of Majorana spinors they are given by $\Xi = \Xi^M \partial_M$ with

$$\Xi^m(x,\Theta) = a^m + i\,\bar{\xi}\gamma^m\Theta + \lambda^m{}_n x^n \qquad (\lambda^{mn} = -\lambda^{nm})$$

$$\Xi^\mu(x,\Theta) = \xi^\mu + \frac{i}{4}\lambda^{mn}(\gamma_{mn}\Theta)^\mu . \tag{1.76}$$

Here the real parameters a^m, $\lambda^m{}_n$ and the Majorana spinor ξ do not depend on the superspace coordinates x, Θ. As pointed out in chapter I.5 the condition $L_\xi E^A = 0$ describes the translations and susy transformations.

Finally we remark that up to numerical factors the superspace metric (1.71) can be directly determined from the requirement of invariance under the transformations (1.76)[9),42)].

(iv) Description of SYM-theories

In the following we briefly indicate how the usual formulation of the pure YM-action on a Riemannian manifold leads to the well-known action for SYM-theories (see part II).

The metric tensorfield g of a Riemannian manifold (M,g) induces a metric $\tilde{g}$ on the space of differential forms defined on M[21)]: if $k(T^r,T^s) = \mathrm{Tr}(T^rT^s)$ represents a well-defined metric on the Lie algebra of the symmetry group of the theory, the action for a YM-field A with curvature 2-form F (defined locally on M) is given by

$$S(A) = \|F\|^2 = \int_M dx\, (\tilde{g}\cdot k)(F, F)$$
$$= \int_M dx\, g^{mp} g^{nq}\, \mathrm{Tr}(F_{mn} F_{pq}) \quad . \tag{1.77}$$

In terms of a local frame field $e^a{}_m(x)$ satisfying

$$g^{mn} = e^m{}_a\, e^n{}_b\, \eta^{ab}$$

this expression can be rewritten as

$$S(A) = \int_M dx\, \mathrm{Tr}(F_{ab} F^{ab}) \quad .$$

Accordingly it can be generalized to our Riemannian superspace by

$$S(\mathbb{A}) = \int d^4x \int d^2\theta \int d^2\bar{\theta}\, \mathrm{Tr}(\mathbb{F}_{AB}\, \mathbb{F}^{AB})$$

where $\mathbb{F} = d\mathbb{A} + \mathbb{A}\mathbb{A} = \frac{1}{2} E^A E^B \mathbb{F}_{BA}$ represents the supercurvature 2-form. Now one has to take into account the solution of the constraints $\mathbb{F}_{\underline{\alpha\beta}} = 0$ and of the reality condition which are generally imposed on the superconnection $\mathbb{A}$:

$$\mathbb{F}_{a\alpha} = i\,(\sigma_a)_{\alpha\dot\beta}\,\overline{W}^{\dot\beta} \qquad \text{with} \qquad \mathcal{D}_\alpha \overline{W}^{\dot\beta} = 0$$

$$\mathbb{F}_{a\dot\alpha} = i\,W^\beta\,(\sigma^a)_{\beta\dot\alpha} \qquad \text{with} \qquad \overline{\mathcal{D}}_{\dot\alpha} W_\beta = 0\ , \quad \mathcal{D}W = \overline{\mathcal{D}}\,\overline{W}$$

$$\mathbb{F}_{ab} = \text{function of } \mathcal{D}_\alpha W_\beta \ \text{ and } \ \overline{\mathcal{D}}_{\dot\alpha}\overline{W}_{\dot\beta}\ . \tag{1.78}$$

Since $\mathrm{Tr}(\mathbb{F}_{a\alpha}\mathbb{F}^{a\alpha})$ is supergauge invariant and $\mathcal{D}_\beta \mathbb{F}_{a\alpha} = 0$ we have $\int d^2\theta\ \mathrm{Tr}(\mathbb{F}_{a\alpha}\mathbb{F}^{a\alpha}) = 0$ (similarly : $\int d\bar\theta^2\ \mathrm{Tr}(\mathbb{F}_{a\dot\alpha}\mathbb{F}^{a\dot\alpha}) = 0$); thus from these geometric considerations one is more or less directly led to postulate the following <u>action for SYM-theories</u>:

$$\begin{aligned} S &= -\frac{1}{4}\int d^4x \int d^2\theta\ \mathrm{Tr}\left(\mathbb{F}^{a\alpha}\,\mathbb{F}_{a\alpha}\right) + h.c. \\ &= \int d^4x \int d^2\theta\ \mathrm{Tr}\left(W^\alpha W_\alpha\right) + h.c. \end{aligned} \tag{1.79}$$

I.11 <u>Miscellaneous facts</u>

In this chapter we have gathered several remarks concerning the generalization of the previously presented formalism and concerning the superspace formulation of supersymmetric field theories. The latter discussion will be continued in parts II, III and IV with the study of SYM-theories.

(i) <u>Relation between the structure of superspace and constraints in susy theories</u>

In the following we would like to indicate how the particular structure of superspace leads to the consideration of certain susy constraints. We recall that such constraints have to be imposed on <u>general superfields</u>

$$\begin{aligned} \phi(x,\theta,\bar\theta) &= \varphi(x) + \theta\psi(x) + \bar\theta\bar\chi(x) \\ &\quad + \theta\theta M(x) + \bar\theta\bar\theta N(x) + (\theta\sigma^m\bar\theta)\,v_m(x) \\ &\quad + \theta\theta\,\bar\theta\bar\lambda(x) + \bar\theta\bar\theta\,\theta\eta(x) + \theta\theta\,\bar\theta\bar\theta\,D(x) \end{aligned} \tag{1.80}$$

(with complex-valued component fields) in order to obtain <u>irreducible representations of the susy algebra</u>; in fact different subsets of component fields of $\phi(x,\theta,\bar\theta)$ transform among themselves and define such irreducible representations (see e.g. section 4.3 of ref. 128).

The natural parametrization of the SP-group,

$$G(x,\theta,\bar\theta;\lambda) = \exp(-ixP + i\theta Q + i\bar\theta\bar Q)\exp(\tfrac{i}{2}\lambda M)$$

$$\equiv G(x,\theta,\bar\theta)\exp(\tfrac{i}{2}\lambda M)$$

leads to a parametrization of rigid superspace in terms of the coordinates $(x^m,\theta^\alpha,\bar\theta_{\dot\alpha})$. Here the fermionic variables $(\theta^\alpha,\bar\theta_{\dot\alpha})$ transform under the reducible representation (1.41) of $SL(2,\mathbb{C})$; thanks to this fact one can find a different parametrization $(y^m,\theta^\alpha,\bar\theta_{\dot\alpha})$ of superspace which allows to impose in a supersymmetric way the constraint that $\phi(y,\theta,\bar\theta)$ does not depend on $\bar\theta$ *). Indeed, in terms of the so-called chiral parametrization[50),128)]

$$G_c(x,\theta,\bar\theta) \equiv e^{-ixP}\, e^{i\bar\theta\bar Q}\, e^{i\theta Q}$$
$$= G\big(x + i(\theta\sigma^m\bar\theta),\theta,\bar\theta\big) \tag{1.81}$$

the left-invariant operators D_α, $\bar D_{\dot\alpha}$ are given by

$$D_\alpha = \frac{\partial}{\partial\theta^\alpha} + 2i(\sigma^m\bar\theta)_\alpha\frac{\partial}{\partial y^m} \qquad (y^m \equiv x^m + i\theta\sigma^m\bar\theta)$$
$$\bar D_{\dot\alpha} = -\frac{\partial}{\partial\bar\theta^{\dot\alpha}} \tag{1.82}$$

and the (supersymmetric) constraint

$$0 = \bar D_{\dot\alpha}\phi(y,\theta,\bar\theta) = -\frac{\partial}{\partial\bar\theta^{\dot\alpha}}\phi(y,\theta,\bar\theta)$$

defines an irreducible representation of supersymmetry, namely the chiral superfield:

$$\phi(y,\theta) = A(y) + \sqrt{2}\,\theta\Psi(y) + \theta\theta F(y). \tag{1.83}$$

Analogously, for the anti-chiral parametrization

$$G_a(x,\theta,\bar\theta) \equiv e^{-ixP}\, e^{i\theta Q}\, e^{i\bar\theta\bar Q}$$
$$= G\big(x - i(\theta\sigma^m\bar\theta),\theta,\bar\theta\big) \tag{1.84}$$

---

*) Note that for the natural coordinates $(x,\theta,\bar\theta)$ the "naïve" constraint $\partial/\partial\bar\theta^{\dot\alpha}\,\phi(x,\theta,\bar\theta) = 0$ is not a supersymmetric one, since the linear operator $\partial/\partial\bar\theta^{\dot\alpha}$ is not invariant under susy transformations.

one has

$$D_\alpha = \frac{\partial}{\partial\theta^\alpha} \tag{1.85}$$

$$\bar{D}_{\dot\alpha} = -\frac{\partial}{\partial\bar\theta^{\dot\alpha}} - 2i\,\theta^\alpha\sigma^m_{\alpha\dot\alpha}\frac{\partial}{\partial y^{+m}} \qquad (y^{+m} \equiv x^m - i\theta\sigma^m\bar\theta)$$

and the differential constraint

$$0 = D_\alpha\phi(y^+,\theta,\bar\theta) = \frac{\partial}{\partial\theta^\alpha}\phi(y^+,\theta,\bar\theta)$$

defines the anti-chiral superfield.

With respect to the chiral parametrization the invariant frame $(E_A{}^M)$ takes a "diagonal" form,

$$\left(E_A{}^M(y,\theta,\bar\theta)\right) = \left[\begin{array}{cc|c} E_a{}^m = \delta_a{}^m & E_a{}^\mu = 0 & 0 \\ E_\alpha{}^m = 2i(\sigma^m\bar\theta)_\alpha & E_\alpha{}^\mu = \delta_\alpha{}^\mu & 0 \\ \hline 0 & 0 & E^{\dot\alpha}{}_{\dot\mu} = \delta^{\dot\alpha}{}_{\dot\mu} \end{array}\right]$$

and so does the corresponding superspace metric,

$$\left(g_{MN}(y,\theta,\bar\theta)\right) = \left[\begin{array}{cc|c} g_{mn} = \eta_{mn} & g_{m\nu} = 2i(\sigma_m\bar\theta)_\nu & 0 \\ g_{\mu n} = -2i(\sigma_n\bar\theta)_\mu & g_{\mu\nu} = \varepsilon_{\mu\nu}[K + 8(\theta\theta)] & 0 \\ \hline 0 & 0 & g^{\dot\mu\dot\nu} = K\,\varepsilon^{\dot\mu\dot\nu} \end{array}\right]$$

(Analogously for the anti-chiral case.)

(ii) The projection method applied to susy transformations and actions

Instead of defining the components of the superfield ϕ by its θ-expansion (1.80) it is often more convenient to define these fields by projecting the superfield ϕ to space-time with the left-invariant differential operators D_α, $\bar{D}_{\dot\alpha}$:

$$\phi| \;,\; D_\alpha\phi| \;,\; \bar{D}_{\dot\alpha}\phi| \;,\; D_\alpha\bar{D}_{\dot\beta}\phi| \;,\; \ldots \quad . \tag{1.86}$$

Here $\phi| \equiv \phi(x,0,0)$. Since $\{D_\alpha, \bar{D}_{\dot\beta}\} \sim \sigma^m_{\alpha\dot\beta}\,\partial_m$ the so-defined component fields are not all independent of each other and one has to determine the independent ones while taking into account eventual constraints on ϕ. The first few of these space-time fields coincide with those of the θ-expansion, but the higher components differ by space-time derivatives of lower ones ; this corresponds to a field-redefinition without physical significance. (For further details and applications, see ch. 3.6 of ref. 57).)

The susy transformations of the component fields (1.86) can be evaluated by a simple method which uses only the algebra of the invariant operators $D_{\underline{\alpha}}$. In fact the latter are related to the supersymmetry generators $Q_{\underline{\alpha}}$ by $\theta,\bar\theta$-dependent terms, and therefore

$$\left[\left(\zeta Q + \bar\zeta\bar Q\right)F\right]\Big| = \left[\left(\zeta D + \bar\zeta \bar D\right)F\right]\Big| = \left[\xi^A D_A F\right]\Big|$$

for any superfield F ; in the last equality we have considered the susy generating vector field (1.5) and used the fact that its spatial component ξ^a satisfies $\xi^a| = 0$, if one does not incorporate ordinary translations. Now the <u>susy transformations of the components (1.86)</u> are obtained as follows : define

$$\delta_\xi\left(F|\right) := \left(\delta_\xi F\right)\Big| \tag{1.87}$$

for the superfields $F = \phi, D_\alpha\phi, \bar D_{\dot\alpha}\phi, \ldots,$ then

$$\delta_\xi F = L_\xi F = \xi^A D_A F \qquad \text{(with } \xi \text{ given by (1.5))} \tag{1.88}$$

yields the susy transformations of the space-time fields.

For instance the independent components of the chiral superfield ϕ are given by

$$\phi| \equiv A \quad , \quad D_\alpha\phi| \equiv \sqrt{2}\,\psi_\alpha \quad , \quad D^\alpha D_\alpha \phi| \equiv -4F \tag{1.89}$$

and the described procedure immediately leads to the familiar susy transformation laws

$$\begin{aligned} \delta_\zeta A &= \sqrt{2}\,\zeta\psi \\ \delta_\zeta \psi_\alpha &= i\sqrt{2}\,(\sigma^m\bar\zeta)_\alpha\,\partial_m A + \sqrt{2}\,\zeta_\alpha F \\ \delta_\zeta F &= i\sqrt{2}\,\bar\zeta\,\bar\sigma^m(\partial_m\psi) \end{aligned} \tag{1.90}$$

The projection method also allows a straightforward evaluation of products of superfields. For instance the infinitesimal supergauge transformation law of a chiral superfield ϕ coupled to a gauge multiplet is given by (see part II)

$$\delta_g \phi = -i \Lambda \phi$$

where

$$\Lambda(y, \theta) = \hat{A} + \sqrt{2}\, \theta \hat{\psi} + \theta\theta \hat{F}$$

represents a chiral superfield ; now the gauge transformation laws of the components (1.89) directly follow by applying the rules

$$\begin{aligned} \delta_g F &:= -i \Lambda F \\ \delta_g (D_\alpha F) &:= D_\alpha (\delta_g F) \\ \delta_g (F|) &:= (\delta_g F) \end{aligned} \tag{1.91}$$

to the superfields $F = \phi,\ D_\alpha \phi,\ D^2 \phi$:

$$\begin{aligned} \delta_g A &= -i (\hat{A} A) \\ \delta_g \psi &= -i (\hat{A} \psi + \hat{\psi} A) \\ \delta_g F &= -i (\hat{A} F + \hat{F} A - \hat{\psi} \psi) \quad . \end{aligned} \tag{1.92}$$

If the action of a superfield ϕ is given as an integral over superspace - or over parts of it - one can easily obtain an expression in terms of its component fields by replacing the fermionic integrations by spinorial derivatives[154]:

$$\begin{aligned} \int d^4x \int d^4\theta\, F(x, \theta, \bar{\theta}) &= -\frac{1}{4} \int d^4x \int d^2\theta\, (\bar{D}^2 F)\Big|_{\bar{\theta}=0} \\ &= -\frac{1}{4} \int d^4x \int d^2\bar{\theta}\, (D^2 F)\Big|_{\theta=0} \end{aligned} \tag{1.93}$$

E.g. for the components (1.89) of the chiral superfield ϕ we obtain the familiar kinetic energy terms from

$$S = \int d^4x \int d^4\theta\, \phi^\dagger \phi = \frac{1}{16} \int d^4x\, \bar{D}^2 D^2 (\phi^\dagger \phi)\Big|$$

$$= \frac{1}{16} \int d^4x \, \bar{D}^2 (\phi^+ D^2 \phi)|$$

$$= \ldots = \int d^4x \left[i(\partial_m \bar{\psi}) \bar{\sigma}^m \psi + A^*(\Box A) + F^* F \right] .$$

(iii) Rigid supersymmetry on non-trivial base manifolds

It is possible to formulate rigid supersymmetric theories on more general spaces than ordinary Minkowski space (upon removing the degrees of freedom associated with gravity). Such a formulation has recently been given for $\mathbb{R} \times S^3$, i.e. ordinary 3-space replaced by the 3-sphere[123]: by taking the limit of large radius this theory allows to draw conclusions about the supersymmetry breaking in $\mathbb{R}^4$. Although the physical relevance of such theories is not completely clear, they exhibit interesting features. (For instance on $\mathbb{R} \times S^3$ the number of bosonic and fermionic states is not equal at each energy level.) Furthermore they lead to superspaces which have a richer structure than the usual one and which have not yet been explored in detail.

(iv) Remarks on the generalizations to supergravity

In section I.5 (ii) we characterized the rigid susy transformations as the unique transformations which leave the familiar rigid vielbein invariant, $L_\xi E^A = 0$. We remark that the gauged (x-dependent) susy transformations[154),

$$\begin{aligned} \xi^a(z) &= 2i \left[\theta \sigma^a \bar{\zeta}(x) - \zeta(x) \sigma^a \bar{\theta} \right] \\ \xi^\alpha(z) &= \zeta^\alpha(x) \\ \bar{\xi}_{\dot{\alpha}}(z) &= \bar{\zeta}_{\dot{\alpha}}(x) \end{aligned} \tag{1.94}$$

are given by a certain Lorentz-covariantized version of $L_\xi E^A = 0$. This can be seen as follows. We define $\xi^a| \equiv 0$, $\xi^{\underline{\alpha}}| \equiv \zeta^{\underline{\alpha}}$ and consider a general Lorentz algebra valued connection $\phi = dz^M \phi_M$ as well as an arbitrary supervielbein $E^A = dz^M E_M{}^A$. When imposing the usual constraints on the components of the torsion, i.e.

$$T_{\alpha\dot{\beta}}{}^c = T_{\dot{\beta}\alpha}{}^c = 2i\, \sigma_{\alpha\dot{\beta}}{}^c$$

$$0 = T_{ab}{}^c = T_{\underline{\alpha} b}{}^c = T_{\alpha\beta}{}^c = T_{\alpha\beta}{}^c = T_{\dot{\alpha}\dot{\beta}}{}^c = T_{\underline{\alpha}\underline{\beta}}{}^{\underline{\gamma}}$$

and choosing the WZ-supergauge for E^A and ϕ, i.e.

$$\left(E_M{}^A\right)\Big| = \begin{bmatrix} e_m{}^a & \frac{1}{2}\psi_m{}^{\underline{\alpha}} \\ 0 & \delta_\mu{}^{\underline{\alpha}} \end{bmatrix} , \qquad \phi_\mu\Big| = 0 \qquad (1.95)$$

the supergravity transformations (1.94) are determined by the condition

$$\left(\mathcal{L}_\xi E\right)_\mu{}^a\Big| = 0 = \left(\mathcal{L}_\xi E\right)^{\mu a}\Big|$$

where $\mathcal{L}_\xi \equiv i_\xi\mathcal{D} + \mathcal{D}i_\xi$ represents the Lorentz-covariant Lie derivative:

$$\left(\mathcal{L}_\xi E\right)_M{}^A = \mathcal{D}_M \xi^A + E_M{}^B \xi^C T_{CB}{}^A .$$

The chiral superfield (1.83) now becomes a Lorentz covariantly chiral field Φ,

$$\bar{\mathcal{D}}_{\dot\alpha} \Phi = 0 .$$

Its components may be defined by projection in analogy to equation (1.89), but with D_α replaced by the covariant derivative $\mathcal{D}_\alpha$. The gauged susy transformations of these space-time fields follow from the definitions

$$\delta_\xi (F|) := (\delta_\xi F)|$$

and

$$\delta_\xi F = \mathcal{L}_\xi F = i_\xi \mathcal{D} F = \xi^A \mathcal{D}_A F \quad \text{(with } \xi \text{ given by (1.94))}$$

(ref. 154, chapt. XIX).

II. THE GENERAL STRUCTURE OF SYM-THEORIES

In complete analogy to the geometric description of ordinary gauge theories on Minkowski space one can formulate SYM-theories in a geometric way on rigid superspace. However a novel feature is the necessity for imposing certain constraints on the YM-superconnection in order to obtain an interesting field theory upon projection to space-time.

The general theory can be formulated in three different, although essentially equivalent ways ; these are usually called the different representations or basis of SYM-theories. For greater clarity we have based our approach on the real representation which allows an easy introduction of the chiral and the anti-chiral ones and a clear discussion of all the reality properties. The latter are essential for obtaining the proper space-time theory upon projection of the superfields. All the results will be schematically summarized at the end of this part.

The geometric approach to SYM-theories has been proposed and developped by J. Wess and his collaborators[153,65]; with a different emphasis from ours these topics are also discussed in refs 57), 66), 128), 155).

II.1 The original approach and the WZ-supergauge

This chapter*) is intended for motivation and orientation; one of the aims of the systematic study presented in the following chapters will be to recover the formulation given here from a more geometric approach.

Originally[152,51] SYM-theories have been developped in the gauge chiral representation where the matter fields are described by a multiplet ϕ of chiral superfields $\phi_1, \ldots, \phi_N$ whereas the gauge field is represented by a real Lie algebra-valued superfield $V = V^{(r)}T^r$ (T^r = generator of the internal symmetry group); the latter contains a large number of component fields usually denoted by C, χ_α, M, N, v_m, λ_α, D. In this approach supergauge transformations are defined by means of a Lie algebra valued chiral superfield $\Lambda = \Lambda^{(r)}T^r$ ("chiral supergauge group" : $\bar{D}_{\dot\alpha} \Lambda^{(r)} = 0$) in analogy to ordinary gauge transformations:

$$\phi' = e^{-i\Lambda} \phi \quad , \quad (\phi^\dagger)' = \phi^\dagger e^{i\Lambda^\dagger} \tag{2.1a}$$

resp.

$$e^{V'} = e^{-i\Lambda^\dagger} e^{V} e^{i\Lambda} . \tag{2.1b}$$

*) (which is primarily based on the chapters VI and VII of ref. 154)).

The (supersymmetric and supergauge invariant) interaction Lagrangian is given by*)

$$S_\phi = \int d^4x \int d^4\theta \; \phi^+ e^V \phi \quad . \tag{2.2}$$

The real multiplet $V(x,\theta,\bar\theta)$ can be viewed as a supersymmetric generalization of the ordinary gauge field $v_m(x)$. The appropriate superspace generalization of the corresponding field-strength $v_{mn} = \partial_m v_n - \partial_n v_m + \frac{i}{2}[v_m, v_n]$ is given by the chiral superfield

$$\mathcal{W}_\alpha = -\frac{1}{8} \bar{D}^2 e^{-V} D_\alpha e^V \tag{2.3}$$

whose supergauge transformation law follows from (2.1b):

$$\mathcal{W}_\alpha' = e^{-i\Lambda} \mathcal{W}_\alpha e^{i\Lambda} \quad . \tag{2.4}$$

The associated supersymmetric and supergauge invariant action reads

$$S_W = \int d^4x \int d^2\theta \; \mathrm{Tr}(\mathcal{W}^\alpha \mathcal{W}_\alpha) + h.c. \quad . \tag{2.5}$$

Most component fields of V correspond to gauge degrees of freedom and can be eliminated by an appropriate supergauge transformation. A very convenient choice is the so-called <u>Wess-Zumino (WZ) supergauge</u>[152] which is characterized by $V^3 = 0$:

$$V = -(\theta\sigma^m\bar\theta)\, v_m + \left[i(\theta\theta)(\bar\theta\bar\lambda) + h.c.\right] + \frac{1}{2}(\theta\theta)(\bar\theta\bar\theta)\, D \quad . \tag{2.6a}$$

In this gauge the superfield-strength $\mathcal{W}_\alpha$ takes the simple form

$$\begin{aligned}\mathcal{W}_\alpha(y,\theta,\bar\theta) = -\frac{i}{2}\lambda_\alpha + \frac{1}{2}\theta_\beta\left[\delta_\alpha{}^\beta D - i\, v_{mn}\,(\sigma^{mn})_\alpha{}^\beta\right] \\ + \frac{1}{2}(\theta\theta)\,\sigma^m_{\alpha\dot\beta}\,(\mathcal{D}_m\bar\lambda^{\dot\beta})\end{aligned} \tag{2.6b}$$

where $y^m = x^m + i\,\theta\sigma^m\bar\theta$ and $\mathcal{D}_m\bar\lambda^{\dot\beta} = \partial_m\bar\lambda^{\dot\beta} + \frac{i}{2}[v_m, \bar\lambda^{\dot\beta}]$.

The <u>characteristic properties of the WZ-gauge</u> can be summarized as follows.

(1) It amounts to using the supergauge freedom to reduce the number of component fields of V to a minimal one ; these are the physically relevant ones, i.e. the

---

*) Here and in the following we assume that the coupling constant g takes the value 1/2 ; arbitrary values for g can be introduced by the rescaling $V \rightarrow 2gV$.

YM-potential v_m and the gaugino field λ_α, as well as the auxiliary field D which is necessary to close the susy algebra off-shell (up to gauge transformations, see (4) below).

(2) After fixing this supergauge one is left with a residual gauge freedom consisting of the ordinary gauge transformations, i.e. transformations with a real Lie algebra valued parameter $v(x)$; at the infinitesimal level these are given by

$$\delta_v v_m = \mathcal{D}_m v \quad , \quad \delta_v \lambda_\alpha = -\frac{i}{2}[v, \lambda_\alpha] \quad , \quad \delta_v D = -\frac{i}{2}[v, D] \tag{2.7a}$$

for the gauge multiplet and

$$\delta_v A = \frac{i}{2} v A \quad , \quad \delta_v \psi_\alpha = \frac{i}{2} v \psi_\alpha \quad , \quad \delta_v F = \frac{i}{2} v F \tag{2.7b}$$

for the matter multiplet ϕ (with components A, ψ_α, F).

(3) The WZ-gauge choice simplifies considerably all computations involving component fields ; e.g. the (supergauge invariant) Lagrangian (2.2) which is expressed in terms of the gauge dependent superfield V takes a non-exponential form in this gauge and thereby reflects in a clear way the particle content of the theory:

$$\begin{aligned} S_\phi = \int d^4x \Big\{ & A^+ \mathcal{D}^m \mathcal{D}_m A - i \psi \sigma^m \mathcal{D}_m \bar{\psi} + i \frac{\sqrt{2}}{2} \left[A^+ (\lambda \psi) - (\bar{\psi} \bar{\lambda}) A \right] \\ & + \frac{1}{2} A^+ D A + F^+ F \Big\} \end{aligned} \tag{2.8a}$$

where $\mathcal{D}_m A = \partial_m A + \frac{i}{2} v_m A$ (similarly for $\mathcal{D}_m \psi$). Analogously an evaluation of the gauge field action (2.5) in the WZ-gauge (2.6b) leads to the simple expression

$$S_W = \int d^4x \left\{ -\frac{1}{4} v_{mn} v^{mn} - i \lambda \sigma^m (\mathcal{D}_m \bar{\lambda}) + \frac{1}{2} D^2 \right\} . \tag{2.8b}$$

(4) However the WZ-gauge has the unpleasant property of not being supersymmetric : component fields of V vanishing in this gauge acquire a non-zero value by an (infinitesimal) susy transformation, e.g.

$$\delta_\xi \chi_\alpha = i (\sigma^m \bar{\xi})_\alpha v_m + \text{terms in } C, M, N .$$

To preserve the WZ-gauge condition when considering susy transformations one has to perform simultaneously a "compensating" supergauge transformation with a (chiral) parameter Λ which depends in a specific way on the component fields of V[*)]:

$$\Lambda(y,\theta) = \sqrt{2}\,\theta^{\alpha}\left[-\frac{i}{\sqrt{2}}(\sigma^m\bar{\xi})_{\alpha}\, v_m(y)\right] + \theta\theta\left[-\bar{\xi}\bar{\lambda}(y)\right] \tag{2.9}$$

$$\left(y^m = x^m + i\,(\theta\sigma^m\bar{\theta})\,\right) .$$

The resulting transformations of the components of the gauge and matter multiplets are usually called the "susy transformations in the WZ-gauge" and are given by:

$$(\delta_A + \delta_g)\,\{C,\ \chi_{\alpha},\ M+iN\} = 0$$

$$(\delta_A + \delta_g)\, v_m = i\,(\xi\sigma_m\bar{\lambda} + \bar{\xi}\bar{\sigma}_m\lambda)$$

$$(\delta_A + \delta_g)\,\lambda_{\alpha} = (\sigma^{mn}\xi)_{\alpha}\, v_{mn} + i\,\xi_{\alpha} D$$

$$(\delta_A + \delta_g)\, D = -\xi\sigma^m(\mathcal{D}_m\bar{\lambda}) + \bar{\xi}\bar{\sigma}^m(\mathcal{D}_m\lambda)$$

$$(\delta_A + \delta_g)\, A = \sqrt{2}\,\xi\psi$$

$$(\delta_A + \delta_g)\,\psi_{\alpha} = i\sqrt{2}\,(\sigma^m\bar{\xi})_{\alpha}\,\mathcal{D}_m A + \sqrt{2}\,\xi_{\alpha} F$$

$$(\delta_A + \delta_g)\, F = i\sqrt{2}\,\bar{\xi}\bar{\sigma}^m(\mathcal{D}_m\psi) + i\,(\bar{\xi}\bar{\lambda})\,A \tag{2.10}$$

where

$$v_{mn} = \partial_m v_n - \partial_n v_m + \frac{i}{2}\,[v_m, v_n]$$

$$\mathcal{D}_m\lambda = \partial_m\lambda + \frac{i}{2}\,[v_m, \lambda] .$$

These transformations do not really define a representation of the susy algebra, but only a "representation modulo an ordinary gauge transformation with field-dependent parameter α"[152,40,128)]:

$$[\delta_1, \delta_2] = -2i\,(\xi_1\sigma^m\bar{\xi}_2 - \xi_2\sigma^m\bar{\xi}_1)\,\partial_m + \delta_{\alpha}^{gauge}$$

$$\alpha(x) = 2i\,(\xi_1\sigma^m\bar{\xi}_2 - \xi_2\sigma^m\bar{\xi}_1)\, v_m(x) . \tag{2.11}$$

[*)] A more detailed discussion can be found in refs 130), 57). Note that this situation is analogous to the one encountered in electrodynamics when the non Lorentz-covariant Coulomb gauge is chosen and Lorentz transformations are considered.

As we will explain in more detail in section III.2 (ii) this algebra exactly corresponds to the algebra of gauge covariant derivatives of the constrained SYM-theory:

$$[\zeta_1^{\alpha} \mathcal{D}_{\alpha} , \zeta_2^{\beta} \mathcal{D}_{\beta}] \phi = -2i(\zeta_1 \sigma^m \bar{\zeta}_2 - \zeta_2 \sigma^m \bar{\zeta}_1) \mathcal{D}_m \phi .$$

From the mathematical point of view these "field-dependent algebras" are obscure objects ; in order to get a better geometric understanding of these objects we will reconsider in part IV the algebra (2.10) from a different point of view and analyse the structure of the associated BRS differential algebra.

We finish these introductory remarks by noting that the WZ-gauge of SYM-theories probably represents the simplest case where field-dependent transformations and algebras occur in supersymmetry and that supergravity offers many other and more subtle examples of these exotic structures. For instance the characteristic properties of the WZ-gauge (1.95) of supergravity are very similar to those summarized above : in this case the residual gauge freedom includes the x-dependent susy transformations whereas the compensating field dependent supergauge transformations are associated with the Lorentz group.

II.2 About formal gauge transformations

In the following we briefly discuss the concept and characteristics of "formal gauge transformations" in ordinary space-time ; such transformations will be considered for the formulation of one of the constraints of SYM-theories.

That two connection forms $A, \tilde{A}$ are related by a formal gauge transformation means that the transformation relating these quantities has exactly the form of a gauge transformation,

$$\tilde{A} = X^{-1} A X + X^{-1} dX ,$$

but that both quantities actually transform under different gauge groups, i.e. the corresponding gauge functions have different properties. A familiar example for this type of transformations has already been mentionned in section I.9 (iv) : the connection form $\phi^a{}_b$ representing locally on a manifold M a given linear connection and the Christoffel symbols $(\Gamma_\ell)^m{}_n$ describing locally a corresponding covariant derivative are related by a formal gauge transformation given by the vielbein fields $e^a{}_m(x)$ (see eq. (1.67)). In this example we point out the following : if the manifold M is endowed with a Riemannian metric g and if the vector fields $D_a = e^m{}_a \partial_m$ (which are the duals of the vielbein forms $e^a = e^a{}_m dx^m$) are chosen to be orthogonal with respect to g, the indices a and m transform with different groups, namely the orthogonal and the general linear groups:

$$(e^a{}_n)' = (O\,e\,L)^a{}_n = O^a{}_b\; e^b{}_m\; L^m{}_n \tag{2.12}$$

where $O \in O(n,\mathbb{R})$ and $L \in GL(n,\mathbb{R})$. We will see below that this also applies to the prepotentials $\mathcal{U},\mathcal{V}$ which describe the formal gauge transformations relating the YM-superconnections of the real, chiral and antichiral representations (eqs (2.35),(2.52)).

II.3 Geometric framework of SYM-theories[154)]

(i) Generalities

As internal symmetry group we consider a compact matrix Lie group G with Lie algebra $\mathfrak{g}$ and suppose that its (hermitean) generators T_r are normalized by

$$\mathrm{Tr}\,(T_r\,T_s) = \delta_{rs} \qquad (r,s \in \{1,\dots,\dim G\}) \; .$$

The corresponding structure constants f_{qrs} are given by

$$[\,T_r\,,\,T_s\,] = i\,f_{qrs}\,T_q$$

and are assumed to be totally antisymmetric. In the adjoint representation of G the generators T_r are represented by $(T_q)_{rs} = -i\,f_{qrs}$.

To formulate YM-theories on superspace $M = SP/L$ in a manifestly supersymmetric way we use the geometric set-up presented in part I ; in particular super p-forms are expressed with respect to the left-invariant basis (E^A) of T_z^*M :

$$Q = \frac{1}{p!}\,E^{A_1}\dots E^{A_p}\,Q_{A_p\dots A_1}$$

where $Q_{A_p\dots A_1}$ is graded antisymmetric in all its indices.

The basic field is the YM-superconnection (gauge potential), i.e. a $\mathfrak{g}$-valued 1-form on M,

$$\begin{aligned} \mathbb{A} &= E^A\,\mathbb{A}_A \equiv E^a\,\mathbb{A}_a + E^\alpha\,\mathbb{A}_\alpha + E_{\dot\alpha}\,\mathbb{A}^{\dot\alpha} \\ \mathbb{A}_A(z) &= \mathbb{A}_A^{(r)}(z)\,T_r \; . \end{aligned} \tag{2.13}$$

transforming under supergauge transformations

$$\mathcal{X} : M \longrightarrow G$$
$$z \longmapsto \mathcal{X}(z) \tag{2.14a}$$

as

$$\mathbb{A}' \equiv {}^{\mathcal{X}}\mathbb{A} = \mathcal{X}^{-1} \mathbb{A} \mathcal{X} - \mathcal{X}^{-1} d\mathcal{X} \tag{2.15}$$

where T_r is in the adjoint representation of $\mathfrak{g}$.

Since our symmetry group G is assumed to be compact, we can rewrite $\mathcal{X}(z)$ as (ref. 61), chap. VI.2)

$$\mathcal{X}(z) = \exp \left[T_r \, \mathfrak{X}^{(r)}(z) \right] \quad . \tag{2.14b}$$

The <u>curvature</u> (field-strength) associated to $\mathbb{A}$ is the $\mathfrak{g}$-valued 2-form

$$\mathbb{F} = d\mathbb{A} + \mathbb{A}\mathbb{A} = \frac{1}{2} E^A E^B \mathbb{F}_{BA} \tag{2.16}$$

with the gauge transformation law

$$\mathbb{F}' \equiv {}^{\mathcal{X}}\mathbb{F} = \mathcal{X}^{-1} \mathbb{F} \mathcal{X} \qquad (T_r \text{ in the adj. repres.}) \tag{2.17}$$

and components given by $(T_{BA}{}^C \equiv T_{BA}{}^C [\phi=0])$

$$\mathbb{F}_{BA} = D_B \mathbb{A}_A - (-)^{ab} D_A \mathbb{A}_B - [\mathbb{A}_B , \mathbb{A}_A] + T_{BA}{}^C \mathbb{A}_C \quad . \tag{2.18}$$

Here and in the following we always suppose that the commutator of two $\mathfrak{g}$-valued superforms P_A, Q_B (of respective form degrees p,q and with susy algebra indices A,B) is bigraded by the form and by the statistics degree:

$$[P_A , Q_B] := P_A Q_B - (-)^{pq+ab} Q_B P_A \quad ; \tag{2.19}$$

in particular the commutator of two $\mathfrak{g}$-valued superfields $\mathbb{A}_A$, $\mathbb{A}_B$ is given by

$$[\mathbb{A}_A , \mathbb{A}_B] = \mathbb{A}_A \mathbb{A}_B - (-)^{ab} \mathbb{A}_B \mathbb{A}_A \; .$$

From part I we recall that the unusual linear term $T_{BA}{}^C \mathbb{A}_C$ in (2.18) is due to the fact that (E^A) represents an anholonomic frame (see (1.30), (1.31) and (1.63)).

We now define the YM-covariant derivative of Lie algebra valued q-forms Q (transforming tensorially under the gauge group) by

$$\mathcal{D}^{(A)} Q := dQ - (-)^q [A, Q] = dQ + [Q, A] \tag{2.20}$$

and note that it transforms under (2.14) as

$$\mathcal{D}^{(A')} = X^{-1} \mathcal{D}^{(A)} X \quad . \tag{2.21}$$

As we will consider different superconnections in the following it is important to indicate explicitly the connection with respect to which the covariant derivative is defined.

As a consequence of its very definition the field-strength $\mathbb{F}$ satisfies the Bianchi identity

$$0 = \mathcal{D}^{(A)} \mathbb{F} = d\mathbb{F} - \mathbb{A}\mathbb{F} + \mathbb{F}\mathbb{A} \tag{2.22a}$$

which can be expressed in terms of the component superfields (2.18) as

$$\sum_{\substack{\text{graded cyclic} \\ \text{permutations} \\ \text{of } (A,B,C)}} \left(\mathcal{D}_C^{(A)} \mathbb{F}_{BA} + T_{CB}{}^{D} \mathbb{F}_{DA} \right) = 0 \quad . \tag{2.22b}$$

Matter fields are described by multiplets of superfields Φ_i whose precise definition depends on the representation chosen for $\mathbb{A}$; the YM-covariant derivative of Φ is defined by

$$\mathcal{D}^{(A)} \Phi = (d - \mathbb{A}) \Phi \qquad \text{for} \quad \Phi = \begin{pmatrix} \Phi_1 \\ \vdots \\ \Phi_N \end{pmatrix} \quad . \tag{2.23}$$

(Here the group generators T_r are supposed to be in a N-dimensional representation of the Lie algebra $\mathfrak{g}$). The multiplets Φ and $\Phi^{\dagger}$ transform under (2.14) as

$$\Phi' \equiv {}^{X}\Phi = X^{-1} \Phi \qquad \text{resp.} \quad (\Phi^{\dagger})' \equiv {}^{X}(\Phi^{\dagger}) = \Phi^{\dagger} (X^{-1})^{\dagger} \tag{2.24}$$

which implies the familiar transformation law

$$\mathcal{D}^{(A')} \Phi' = X^{-1} (\mathcal{D}^{(A)} \Phi) \quad .$$

While the (graded) commutator of the ordinary derivatives D_A is given by

$$[D_A , D_B] = - T_{AB}{}^C D_C \quad , \tag{2.25}$$

it follows from the definitions (2.20) and (2.23) that the (graded) commutator of YM-covariant derivatives is expressed by

$$[\mathcal{D}_A^{(A)} , \mathcal{D}_B^{(A)}] Q = - [\mathbb{F}_{AB} , Q] - T_{AB}{}^C \mathcal{D}_C^{(A)} Q \tag{2.26a}$$

and

$$[\mathcal{D}_A^{(A)} , \mathcal{D}_B^{(A)}] \Phi = - \mathbb{F}_{AB} \Phi - T_{AB}{}^C \mathcal{D}_C^{(A)} \Phi \tag{2.26b}$$

for the $\mathcal{G}$-valued fields Q and the matter fields Φ respectively. (Strictly speaking the covariant derivatives $\mathcal{D}_A$ are defined by

$$\mathcal{D}_A = E_A{}^M (\partial_M - \mathbb{A}_M + \phi_M) \quad ,$$

their commutator being

$$[\mathcal{D}_A , \mathcal{D}_B] V_D = - \mathbb{F}_{AB} V_D - R[\phi]_{ABD}{}^C V_C - T[\phi]_{AB}{}^C \mathcal{D}_C V_D \quad ,$$

but for the canonical linear connection $\phi_M \equiv 0$ these expressions obviously reduce to the previously given ones, if we write $T_{AB}{}^C \equiv T[\phi=0]_{AB}{}^C$. This explains the occurence of the unusual T-terms in the equations (2.26)).

We note that our definitions (2.23), (2.24) differ slightly from those of Wess and Bagger[154], but otherwise we consistently use their conventions.

(ii) The constraints on the superconnection

The set-up presented in the previous section is just an obvious superspace generalization of ordinary YM-theories. The theory obtained in this way is characterized by a large number of degrees of freedom (e.g. many space-time fields are contained in the superfields $\mathbb{A}_A$, $\mathbb{F}_{AB}$). In order to reduce this number and obtain an appropriate supersymmetric version of ordinary gauge theories after projection to space-time, one has to impose constraint equations and an adequate reality condition on the superconnection form $\mathbb{A}$.

There are two types of constraints which are usually imposed on the components of the curvature $\mathbb{F}$ associated to $\mathbb{A}$ and which are called the "representation preserving" and the "conventional" constraints respectively. If one wants to allow covariantly chiral and anti-chiral matter superfields in the theory,

$$0 = \bar{\mathbb{D}}^{(A)}_{\dot\alpha} \Phi = (\bar{D}_{\dot\alpha} - \mathbb{A}_{\dot\alpha})\Phi \quad , \quad 0 = \mathbb{D}^{(A)}_{\alpha} \Phi^{\dagger} = D_{\alpha}\Phi^{\dagger} - \Phi^{\dagger}\mathbb{A}_{\alpha} \quad ,$$

then the equations (2.26b), i.e. more precisely

$$[\bar{\mathbb{D}}^{(A)}_{\dot\alpha} , \bar{\mathbb{D}}^{(A)}_{\dot\beta}]\Phi = - \mathbb{F}_{\dot\alpha\dot\beta}\Phi \quad , \quad [\mathbb{D}^{(A)}_{\alpha} , \mathbb{D}^{(A)}_{\beta}]\Phi^{\dagger} = - \Phi^{\dagger}\mathbb{F}_{\alpha\beta}$$

require the following integrability conditions ("<u>representation preserving constraints</u>"):

$$\mathbb{F}_{\dot\alpha\dot\beta} = 0 = \mathbb{F}_{\alpha\beta} \quad . \tag{2.27a}$$

In view of the particular form of $\mathbb{F}_{\alpha\dot\beta}$ and $\mathbb{F}_{\dot\beta\alpha}$,

$$\mathbb{F}_{\dot\beta\alpha} = \mathbb{F}_{\alpha\dot\beta} = D_{\alpha}\mathbb{A}_{\dot\beta} + \bar{D}_{\dot\beta}\mathbb{A}_{\alpha} - [\mathbb{A}_{\alpha}, \mathbb{A}_{\dot\beta}] + 2i\,\sigma^{a}_{\alpha\dot\beta}\,\mathbb{A}_{a} \quad , \tag{2.28}$$

the "<u>conventional constraint</u>"

$$\mathbb{F}_{\alpha\dot\beta} = 0 \tag{2.27b}$$

just amounts to a redefinition of $\mathbb{A}_a$ in terms of $\mathbb{A}_\alpha$, $\mathbb{A}_{\dot\alpha}$ and their spinorial derivatives.

The appropriate <u>reality condition</u> turns out to be the following:

$$\mathbb{A}^{\dagger} \cong -\mathbb{A} \quad . \tag{2.29}$$

Here the cross denotes hermitean conjugation and $\cong$ means that these quantities are related by a formal gauge transformation. In general the latter can only be properly specified after solving explicitly the constraints (2.27a) in terms of unconstrained prepotentials. In the abelian case where (2.15) becomes ${}^{X}\mathbb{A} = \mathbb{A} - d\mathfrak{X}$ the reality condition (2.29) reduces to

Re $\mathbb{A}$ = pure supergauge

which implies that all the relevant information about $\mathbb{A}$ is actually contained in the imaginary part of $\mathbb{A}$.

As a consequence of the constraints (2.27) the Bianchi identities (2.22b) are no more trivially satisfied, rather they define a system of differential equations for the non-constrained superfields $\mathbb{F}_{ab}$, $\mathbb{F}_{a\underline{\alpha}}$. The general solution of this set of equations will be summarized in the next section.

Alternatively one can solve explicitly the constraints (2.27) in terms of unconstrained superfields which will be the object of section (iv).

We will present both solutions without assuming that the YM-superconnection satisfies specific reality properties: the implementation of the reality condition (2.29) will only be considered in a second step (section (v) and following chapters).

We briefly mention that the choice of a connection on superspace which is constrained by the conditions (2.27) is equivalent to the consideration of an unconstrained connection on a certain (super) twistor space (ref. 158), see also ref. 121)) ; although this point of view sheds some light on the general mathematical structure of the theory, it does not seem to be very helpful for a derivation or understanding of concrete space-time results and therefore we will not consider it here.

(iii) Representation independent solution of the Bianchi identities

The general solution of the Bianchi identities (2.22b) subject to the constraints (2.27) is given by[154)]

$$
\begin{aligned}
\mathbb{F}_{a\alpha} &= i\,(\sigma_a)_{\alpha\dot\beta}\,\bar W_1^{\dot\beta} \\
\mathbb{F}_{a\dot\alpha} &= i\,W_2^{\beta}\,(\sigma_a)_{\beta\dot\alpha} \\
\mathbb{F}_{ab} &= \tfrac{1}{2}\left(\mathcal{D}\,\sigma_{ab}\,W_2 - \bar{\mathcal{D}}\,\bar\sigma_{ab}\,\bar W_1\right)
\end{aligned}
\qquad (2.30a)
$$

where W_2^{α}, $\bar W_1^{\dot\alpha}$ are two Lie algebra valued superfields satisfying

$$
\begin{aligned}
\bar{\mathcal{D}}^{(A)}_{\dot\alpha} W_2^{\alpha} &= 0 = \mathcal{D}^{(A)}_{\alpha}\,\bar W_1^{\dot\alpha} \\
\mathcal{D}^{(A)} W_2 &= \bar{\mathcal{D}}^{(A)}\bar W_1 \quad .
\end{aligned}
\qquad (2.30b)
$$

These fields transform as spinors under the Lorentz group and like the curvature $\mathbb{F}$ under supergauge transformations:

$$
(W_2^{\alpha})' = X^{-1}\,W_2^{\alpha}\,X \quad , \quad (\bar W_1^{\dot\alpha})' = X^{-1}\,\bar W_1^{\dot\alpha}\,X \quad . \qquad (2.31)
$$

Contrary to the general practice we have supplemented $\bar W_1^{\dot\alpha}$ and W_2^{α} with subscripts 1 and 2 to indicate that these superfields are not related in any way by hermitean conjugation as long as we do not impose the reality condition (2.29) on the superconnection $\mathbb{A}$. In the real representation the latter will be implemented in such a way that $(W_2^{\alpha})^{\dagger} = \bar{W}_1^{\dot{\alpha}}$, but in the chiral/antichiral representations this relation only holds up to a specific formal gauge transformation (sect. II.5 (ii)).

(iv) Representation independent solution of the constraints

In the following we solve the constraints $\mathbb{F}_{\underline{\alpha\beta}} = 0$ explicitly, i.e. we express the associated connection components $\mathbb{A}_a$, $\mathbb{A}_{\underline{\alpha}}$ in terms of unconstrained superfields.

The condition

$$0 = \mathbb{F}_{\alpha\beta} = D_\alpha \mathbb{A}_\beta + D_\beta \mathbb{A}_\alpha - [\mathbb{A}_\alpha, \mathbb{A}_\beta]$$

is a Maurer-Cartan like equation and thus solved by a Lie algebra valued superfield $\mathbb{A}_\alpha$ which has the form of a "pure gauge potential":

$$\mathbb{A}_\alpha = -\mathcal{V}^{-1} D_\alpha \mathcal{V} \quad . \tag{2.32a}$$

Here $\mathcal{V}(z) = \exp \nu(z)$ with $\nu(z) = T_s \nu^s(z)$ being an unconstrained $\mathcal{G}$-valued superfield so that

$$\mathbb{A}_\alpha = -T_s \left\{ D_\alpha \nu^s + \frac{i}{2} f^s{}_{pq} (D_\alpha \nu^p) \nu^q + \ldots \right\}$$

actually represents a $\mathcal{G}$-valued superfield. The superfield $\mathcal{V}$ (or the equivalent field ν) is usually referred to as prepotential.

Similarly the equation

$$0 = \mathbb{F}_{\dot{\alpha}\dot{\beta}} = \bar{D}_{\dot{\alpha}} \mathbb{A}_{\dot{\beta}} + \bar{D}_{\dot{\beta}} \mathbb{A}_{\dot{\alpha}} - [\mathbb{A}_{\dot{\alpha}}, \mathbb{A}_{\dot{\beta}}]$$

can be solved in terms of a prepotential $\mathcal{U} = \exp \mu$:

$$\mathbb{A}_{\dot{\alpha}} = -\mathcal{U}^{-1} \bar{D}_{\dot{\alpha}} \mathcal{U} \quad . \tag{2.32b}$$

As follows from (2.28) the conventional constraint $\mathbb{F}_{\alpha\dot{\beta}} = 0$ is equivalent to the following redefinition of $\mathbb{A}_a$:

$$\mathbb{A}_a = -\frac{i}{4} \left(D^\alpha \sigma^a_{\alpha\dot{\alpha}} \mathbb{A}^{\dot{\alpha}} + \bar{D}_{\dot{\alpha}} (\bar{\sigma}^a)^{\dot{\alpha}\alpha} \mathbb{A}_\alpha - \{ \mathbb{A}^\alpha, \sigma^a_{\alpha\dot{\alpha}} \mathbb{A}^{\dot{\alpha}} \} \right) \tag{2.33}$$

By using (2.32) this expression can be written in terms of the prepotentials $\mathcal{U}, \mathcal{V}$ and their derivatives.

Thus we have obtained a complete solution of the constraints $\mathbb{F}_{\underline{\alpha\beta}} = 0$ in terms of two unconstrained and unrelated superfields $\mathcal{U}, \mathcal{V}$.

As may be directly verified the transformation laws

$$\begin{aligned} \mathcal{U} &\longrightarrow \mathcal{U}\,X \\ \mathcal{V} &\longrightarrow \mathcal{V}\,X \end{aligned} \tag{2.34a}$$

induce the correct supergauge transformation laws (2.15) of $\mathbb{A}_a$ and $\mathbb{A}_{\underline{\alpha}}$. In addition to the transformations (2.34a) one can always perform so-called pregauge transformations Σ (resp. π) of $\mathcal{U}$ (resp. $\mathcal{V}$) which leave the gauge potentials $\mathbb{A}_a, \mathbb{A}_{\underline{\alpha}}$ invariant:

$$\begin{aligned} \mathcal{U} &\longrightarrow \Sigma_1^{-1}\,\mathcal{U} \qquad (\bar{D}_{\dot\alpha}\Sigma_1 = 0) \\ \mathcal{V} &\longrightarrow \pi^{-1}\,\mathcal{V} \qquad (D_\alpha \pi = 0) \quad . \end{aligned} \tag{2.34b}$$

Here $\Sigma = e^{i\Lambda}$ with a $\mathfrak{g}$-valued chiral superfield Λ and $\pi = e^{i\tilde{\Lambda}}$ with a $\mathfrak{g}$-valued anti-chiral field $\tilde{\Lambda}$. Consequently the complete transformation laws of the prepotentials are postulated to be

$$\begin{aligned} \mathcal{U}' &= \Sigma_1^{-1}\,\mathcal{U}\,X \qquad (\bar{D}_{\dot\alpha}\Sigma_1 = 0) \\ \mathcal{V}' &= \pi^{-1}\,\mathcal{V}\,X \qquad (D_\alpha \pi = 0) \quad . \end{aligned} \tag{2.35}$$

For the general formulation of the theory it is convenient to introduce another prepotential which is defined by

$$\mathcal{W} = \mathcal{V}\cdot\mathcal{U}^{-1} \quad . \tag{2.36}$$

According to (2.35) it transforms as

$$\mathcal{W}' = \pi^{-1}\,\mathcal{W}\,\Sigma_1 \quad . \tag{2.37}$$

(v) The different representations

The three types of superfields which are generally considered in simple supersymmetry are the real, chiral and anti-chiral ones. The different superconnections whose infinitesimal supergauge transformation parameter $\mathfrak{X}$ is given by these superfields are said to define, respectively, the (gauge) real, chiral and anti-chiral representations (or basis) of SYM-theories*).

These three representations correspond to different implementations of the reality condition (2.29) : the real one corresponds to the choice of a superconnection $\mathbb{A}$ which satisfies the strong reality condition

$$\mathbb{A}^{\dagger} = -\mathbb{A} \tag{2.38}$$

whereas the connections characterizing the other representations satisfy in a specific way the weaker condition $\mathbb{A}^{\dagger} \cong -\mathbb{A}$.

After discussing in detail the real representation in the next chapter, we will pass over to the two other ones and study more closely their reality properties.

II.4 The gauge real representation

(i) Reality properties of the component and superfields

According to the rules for complex conjugation outlined in appendix 2, we have the relations

$$(E_a)^* = E_a \quad , \quad (E_\alpha)^* = E_{\dot\alpha} \tag{2.39}$$

which imply that the reality condition $\mathbb{A}^{\dagger} = -\mathbb{A}$ characterizing the gauge real representation is equivalent to

$$\mathbb{A}_a^{\dagger} = -\mathbb{A}_a \quad , \quad \mathbb{A}_\alpha^{\dagger} = -\mathbb{A}_{\dot\alpha} \quad . \tag{2.40}$$

(Note that the operations $*$ and $\dagger$ reverse the order of all products including exterior products.) From these relations it follows that the components (2.18) of the curvature form $\mathbb{F}$ satisfy

$$\mathbb{F}_{ab}^{\dagger} = -\mathbb{F}_{ab} \quad , \quad \mathbb{F}_{a\alpha}^{\dagger} = -\mathbb{F}_{a\dot\alpha} \quad , \quad \mathbb{F}_{\alpha\beta}^{\dagger} = \mathbb{F}_{\dot\alpha\dot\beta} \quad , \quad \mathbb{F}_{\alpha\dot\beta}^{\dagger} = \mathbb{F}_{\dot\alpha\beta} \tag{2.41a}$$

whence the equation

$$\mathbb{F}^{\dagger} = \mathbb{F} \quad . \tag{2.41b}$$

---

*) Here the terms "representation" and "basis" are not to be understood in their strict mathematical sense, but only in the sense of a "realization".

In particular $F^{\dagger}_{a\alpha} = - F_{a\dot{\alpha}}$ implies that the two superfields W_2^{α}, $\bar{W}_1^{\dot{\alpha}}$ occuring in the general solution (2.30) of the Bianchi identities are related by hermitean conjugation:

$$(W_2^{\alpha})^{\dagger} = \bar{W}_1^{\dot{\alpha}} \qquad \text{i.e.} \qquad W_2^{\alpha} = W_1^{\alpha} \quad .$$

Thus in the real representation the general solution of the Bianchi identities is given by a single $\mathcal{G}$-valued and covariantly chiral superfield W_{α} whose covariant projections can be used to specify the field content of the theory. This will be discussed in part III; there we will also show that the gauge superfield action

$$S_W = \int d^4x \int d^2\theta \; \mathrm{Tr}(W^{\alpha} W_{\alpha}) \quad + \quad h.c. \tag{2.42}$$

leads to a component field expression which is exactly of the same form as (2.8b).

Gauge independence of the conditions (2.40) requires that the supergauge transformations are described by "unitary" superfields, $X^{\dagger} = X^{-1}$. This means that the corresponding infinitesimal transformations are given by $\mathcal{G}$-valued superfields $\mathcal{X}$ belonging to the "real supergauge group":

$$X = e^{\mathcal{X}} \quad , \quad \mathcal{X}^{\dagger} = - \mathcal{X} \quad . \tag{2.43}$$

As we assumed our Lie algebra $\mathcal{G}$ to be represented by hermitean generators, equation (2.40) guarantees that the superfields A_a project down to the real components $v_a = T_r v_a^r(x)$ of a connection 1-form v on space-time which can be identified with the ordinary YM-potential:

$$A_a| = - \frac{i}{2} v_a \qquad (v_a^{\dagger} = v_a) \quad . \tag{2.44a}$$

Similarly we have

$$F_{ab}| = - \frac{i}{2} v_{ab} \qquad (v_{ab}^{\dagger} = v_{ab}) \quad . \tag{2.44b}$$

Furthermore by projection of eq. (2.43) the supergauge freedom is directly reduced to the ordinary gauge freedom corresponding to a real $\mathcal{G}$-valued field $v(x)$:

$$\mathcal{X}| = \frac{i}{2} v \qquad (v^{\dagger} = v) \quad . \tag{2.45}$$

As we will see below this is not true in the (anti-) chiral representation unless one fixes the supergauge freedom by choosing the WZ-gauge.

In the real basis the matter fields are described by covariantly chiral multiplets,

$$\Phi = \begin{pmatrix} \Phi_1 \\ \vdots \\ \Phi_N \end{pmatrix} \quad , \qquad \bar{\mathcal{D}}^{(A)}_{\dot{\alpha}} \Phi = 0 \quad , \tag{2.46a}$$

respectively by covariantly anti-chiral ones (apply (2.40)):

$$\Phi^{\dagger} = (\Phi_1^{*}, \ldots, \Phi_N^{*}) \quad , \quad 0 = \mathcal{D}^{(A)}_{\alpha} \Phi^{\dagger} = D_{\alpha} \Phi^{\dagger} + \Phi^{\dagger} A_{\alpha} \quad . \tag{2.46b}$$

The corresponding interaction Lagrangian reads

$$S_{int} = \int d^4x \int d^4\theta \; \Phi^{\dagger} \Phi \quad ; \tag{2.47}$$

its component field expression is exactly of the same form as the familiar Lagrangian in the WZ-gauge, eq.(2.8a) (see part III).

To complete our discussion we analyse the consequences of the reality condition (2.40) for the superfields occuring in the explicit solution of the constraints (section II.3 (iv)).

The conditions (2.40) imply that the prepotentials $\mathcal{U}, \mathcal{V}$ (resp. μ, ν) occuring in the expressions (2.32) are not anymore independent of each other, but related by the equation

$$\mathcal{U}^{\dagger} = \mathcal{V}^{-1} \quad \Longleftrightarrow \quad \mu^{\dagger} = -\nu \quad . \tag{2.48}$$

Thus the whole theory can be expressed in terms of a single unconstrained $\mathcal{G}$-valued superfield. Nevertheless, for notational and mnemonical reasons we will write down all equations in terms of both fields while keeping in mind their equivalence expressed by the relation (2.48).

For the pregauge transformations (2.34b) we obtain

$$\Pi^{\dagger} = \Sigma^{-1} \quad \Longleftrightarrow \quad \Lambda^{\dagger} = \tilde{\Lambda} \tag{2.49}$$

so that the transformation law (2.37) of the prepotential $\mathcal{W}$ becomes

$$\mathcal{W}' = \Sigma^{\dagger} \mathcal{W} \Sigma = e^{-i\Lambda^{\dagger}} \mathcal{W} e^{i\Lambda} \tag{2.50}$$

where $\mathcal{W}$ is now given by

$$\mathcal{W} = \mathcal{V} \cdot \mathcal{V}^{\dagger} \; .$$

This quantity is real (i.e. $\mathcal{W}^\dagger = \mathcal{W}$) and positive and thus may be written as

$$\mathcal{W} = e^V \tag{2.51}$$

where $V = V^\dagger$ is again a $\mathfrak{g}$-valued superfield. The latter corresponds to the vector multiplet considered in the original approach to SYM-theories; in fact insertion of (2.51) into (2.50) yields the familiar transformation law (2.1b) of V.

The transformation laws (2.35) of the prepotentials $\mathcal{U}, \mathcal{V}$ now have the explicit form

$$\begin{aligned} \mathcal{U}' &= e^{-i\Lambda}\,\mathcal{U}\,e^{\mathfrak{X}} \qquad (\bar{D}_{\dot\alpha}\Lambda = 0\,,\quad \mathfrak{X}^\dagger = -\mathfrak{X}) \\ \mathcal{V}' &= e^{-i\Lambda^\dagger}\,\mathcal{V}\,e^{\mathfrak{X}} \qquad (D_\alpha \Lambda^\dagger = 0\,,\quad \mathfrak{X}^\dagger = -\mathfrak{X})\,. \end{aligned} \tag{2.52}$$

Like the vielbein fields $e^a{}_m(x)$ of Riemannian geometry these fields transform with different groups to the right and to the left (see eq.(2.12)) ; their gauge group is larger than the real gauge group under which the real superconnection $\mathbb{A}$ transforms.

Just as the vielbein fields $e^a{}_m(x)$ can be used to describe a formal gauge transformation on space-time (eq.(1.67)), the prepotentials $\mathcal{U}, \mathcal{V}$ can be used to introduce new superconnections on superspace :

$$\begin{aligned} \varphi &:= \mathcal{U}\,\mathbb{A}\,\mathcal{U}^{-1} - \mathcal{U}^{-1}d\mathcal{U} \\ \tilde\varphi &:= \mathcal{V}\,\mathbb{A}\,\mathcal{V}^{-1} - \mathcal{V}^{-1}d\mathcal{V}\,. \end{aligned} \tag{2.53}$$

As an immediate consequence of (2.52) these connections transform under the chiral and anti-chiral gauge groups respectively ; this will be further pursued in chapter II.5.

(ii) Gauge and pregauge fixing ; the WZ-gauge

For some considerations it is useful to have explicit expressions of $\mathcal{U}$ and $\mathcal{V}$ in terms of the real prepotential V. These can be obtained as follows.

The complex Lie algebra valued superfield $\mu \equiv (\mu_1^{(r)} + i\mu_2^{(r)})T_r$ occuring in the prepotential $\mathcal{U} = e^\mu$ has a natural decomposition into a hermitean and an anti-hermitean part :

$$\mu = \mu_h + \mu_a \qquad (\mu_h^\dagger = \mu_h\,,\quad \mu_a^\dagger = -\mu_a)\,.$$

By an appropriate supergauge transformation $\chi = e^{\mathfrak{X}}$ $(\mathfrak{X}^\dagger = -\mathfrak{X})$ the contribution

μ_a can be eliminated:

$$\mathcal{U} X = \mathcal{U}' = e^{\mu'_h} .$$

Now $\mathcal{U}'$ is a real superfield, $(\mathcal{U}')^\dagger = \mathcal{U}'$, and the relation $e^{-V} = \mathcal{U}\mathcal{U}^\dagger$ becomes

$$e^{-V'} = (\mathcal{U}')^2 = e^{2\mu'_h} \qquad i.e. \qquad \mu'_h = -\tfrac{1}{2} V .$$

In this particular supergauge the prepotentials $\mathcal{U}$ and $\mathcal{V}$ therefore take the special form*)

$$\mathcal{U} = e^{-\frac{1}{2}V} = \mathcal{U}^\dagger \quad , \quad \mathcal{V} = e^{\frac{1}{2}V} = \mathcal{V}^\dagger . \tag{2.54}$$

Further restriction to the WZ-gauge consists in using the remaining pregauge freedom of $V = V^\dagger$ to obtain $V^3 = 0$, i.e. a real superfield of the form (2.6a). In this gauge the lowest components of the prepotentials $\mathcal{U}$, $\mathcal{V}$ and $\mathcal{W}$ are trivial:

$$\mathcal{U}| = \mathcal{V}| = \mathcal{W}| = \mathbb{1} .$$

Furthermore in the abelian case it is easy to verify that - according to the equations (2.33), (2.32), (2.54), (2.6a) - the lowest component of $\mathbf{A}_a$ is then given by

$$\mathbb{A}_a| = -\frac{i}{2} v_a$$

where v_a is the vector field occuring (2.6a).

II.5 The chiral/anti-chiral representations

To distinguish the corresponding variables in the different representations we have denoted all "real quantities" by straight letters (e.g. the superconnection $\mathbf{A}$, the curvature $\mathbf{F}$, the field-strength W_α, ...) and all "chiral quantities" by curly letters (e.g. $\mathcal{G}$, $\mathcal{F}$, $\mathcal{W}_\alpha$,...) ; the "anti-chiral" variables are characterized by an additional "tilde" on the defining symbols.

---

*) Note that they only maintain this form under pregauge transformations, if one simultaneously considers a field-dependent X-transformation (recall $\pi^\dagger = \Sigma^{-1}$ and $\mathcal{U}^\dagger = \mathcal{V}^{-1}$)57):

$$\mathcal{V}' = \Pi^{-1} \mathcal{V} X \qquad \text{with} \qquad X = \mathcal{V}^{-1} \Pi (\Pi^{-1} \mathcal{V}^2 \Sigma)^{1/2}$$

resp. $$\mathcal{U}' = \Sigma^{-1} \mathcal{U} X \qquad \text{with} \qquad X = \mathcal{U}^{-1} \Sigma (\Sigma^{-1} \mathcal{U}^2 \Pi)^{1/2} .$$

(i) Superconnections and curvatures

In the last chapter we introduced the superconnections φ, $\tilde{\varphi}$ of the chiral/anti-chiral representations by using the prepotentials $\mathcal{U}$, $\mathcal{V}$ ($\mathcal{V}^{\dagger} = \mathcal{U}^{-1}$) occuring in the explicit solution of the real representation constraints:

$$
\begin{aligned}
\mathcal{U} \mathbb{A} \mathcal{U}^{-1} - \mathcal{U}^{-1} d\mathcal{U} &=: \varphi \equiv E^{A} \varphi_{A} \\
\mathcal{V} \mathbb{A} \mathcal{V}^{-1} - \mathcal{V}^{-1} d\mathcal{V} &=: \tilde{\varphi} \equiv E^{A} \tilde{\varphi}_{A} \quad .
\end{aligned}
\tag{2.55}
$$

From the transformation laws (2.15),(2.35) of $\mathbb{A}$, $\mathcal{U}$ and $\mathcal{V}$ one immediately derives those of the $\mathfrak{g}$-valued 1-forms φ and $\tilde{\varphi}$:

$$
\begin{aligned}
\varphi' &= \Sigma_1^{-1} \varphi \Sigma_1 - \Sigma_1^{-1} d\Sigma_1 \equiv {}^{\Sigma_1}\varphi \qquad (\bar{D}_{\dot{\alpha}} \Sigma_1 = 0) \\
\tilde{\varphi}' &= \Pi^{-1} \tilde{\varphi} \Pi - \Pi^{-1} d\Pi \equiv {}^{\Pi}\tilde{\varphi} \qquad (D_{\alpha} \Pi = 0 , \; \Pi^{\dagger} = \Sigma_1^{-1})
\end{aligned}
\tag{2.56}
$$

Thus φ and $\tilde{\varphi}$ transform as connections under the chiral and the anti-chiral gauge groups respectively. Their curvatures $\mathcal{F}$ and $\tilde{\mathcal{F}}$ are given by

$$
\begin{aligned}
\mathcal{F} &\equiv \tfrac{1}{2} E^{A} E^{B} \mathcal{F}_{BA} := d\varphi + \varphi\varphi = \mathcal{U} \mathbb{F} \mathcal{U}^{-1} \\
\tilde{\mathcal{F}} &\equiv \tfrac{1}{2} E^{A} E^{B} \tilde{\mathcal{F}}_{BA} := d\tilde{\varphi} + \tilde{\varphi}\tilde{\varphi} = \mathcal{V} \mathbb{F} \mathcal{V}^{-1} \quad .
\end{aligned}
\tag{2.57}
$$

As a direct consequence of their definition (2.55) the superforms φ and $\tilde{\varphi}$ are related by a formal gauge transformation involving the real prepotential $\mathcal{W} = \mathcal{U}\mathcal{V}^{-1}$:

$$
\tilde{\varphi} = \mathcal{W} \varphi \mathcal{W}^{-1} - \mathcal{W} d\mathcal{W}^{-1} \quad (\mathcal{W} = e^{V} , \; V^{\dagger} = V) \quad . \tag{2.58}
$$

For the components of $\mathcal{F}$ and $\tilde{\mathcal{F}}$ we have:

$$
\mathcal{F}_{AB} = \mathcal{U} \mathbb{F}_{AB} \mathcal{U}^{-1} \quad , \quad \tilde{\mathcal{F}}_{AB} = \mathcal{V} \mathbb{F}_{AB} \mathcal{V}^{-1} \quad . \tag{2.59}
$$

Since the connection $\mathbb{A}$ is constrained by the conditions $\mathbb{F}_{\underline{\alpha\beta}} = 0$, φ and $\tilde{\varphi}$ are constrained by $\mathcal{F}_{\underline{\alpha\beta}} = 0$ and $\tilde{\mathcal{F}}_{\underline{\alpha\beta}} = 0$ respectively.

Schematically we can summarize the interrelations between the different basis by the following triangle:

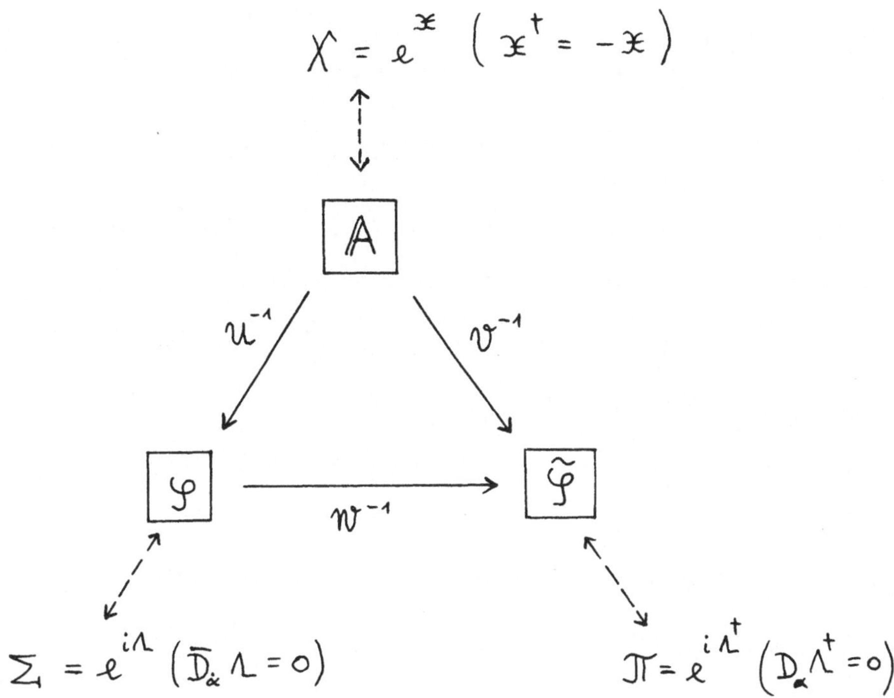

Obviously this triangular structure has its origin in the integrability condition $\mathbf{F}_{\alpha\beta} = 0 = \mathbf{F}_{\dot{\alpha}\beta}$ and in the reality condition $\mathbf{A}_{\alpha}^{\dagger} = -\mathbf{A}_{\dot{\alpha}}$. It is characteristic not only for N=1 SYM-theories, but also for other supersymmetric field theories where similar constraints are imposed, namely for N=1 supergravity*)[57], N=2 SYM-theories in harmonic superspace (ref. 55), see also 38)) and N=2 supergravity in harmonic superspace[39]. (For the last two cases the three equivalent representations are, respectively, the analytic, anti-analytic and a certain real one.)

(ii) <u>Reality properties</u>

By taking the hermitean conjugate of the first of the equations (2.55) and using the reality condition $\mathbf{A}^{\dagger} = -\mathbf{A}$ we obtain

$$\varphi^{\dagger} = -\tilde{\varphi} \qquad . \tag{2.60}$$

Insertion of this relation in (2.58) leads to the <u>reality conditions which caracterize the chiral and anti-chiral representations</u> :

$$-\varphi^{\dagger} = \mathcal{W}\varphi\mathcal{W}^{-1} - \mathcal{W}d\mathcal{W}^{-1} \tag{2.61a}$$

---

*) The algebraic origin of the constraints for N=1 sugra and N=2 SYM are discussed in ref. 58).

resp.

$$-\tilde{\varphi}^{+} = \mathcal{W}^{-1}\tilde{\varphi}\,\mathcal{W} - \mathcal{W}^{-1}d\mathcal{W} \qquad (\mathcal{W} = e^{V}, V^{+} = V) \;. \tag{2.61b}$$

As pointed out before these relations represent a very specific implementation of the weak reality condition (2.29).

From (2.39), (2.57), (2.60), one immediately deduces the following equations:

$$\begin{aligned} &\varphi_a^{+} = -\tilde{\varphi}_a \;, \quad \varphi_\alpha^{+} = -\tilde{\varphi}_{\dot\alpha} \\ &\mathcal{F}^{+} = \tilde{\mathcal{F}} \\ &\mathcal{F}_{ab}^{+} = -\tilde{\mathcal{F}}_{ab} \;, \quad \mathcal{F}_{a\alpha}^{+} = -\tilde{\mathcal{F}}_{a\dot\alpha} \;, \quad \mathcal{F}_{\alpha\beta}^{+} = \tilde{\mathcal{F}}_{\dot\alpha\dot\beta} \;, \quad \mathcal{F}_{\alpha\dot\beta}^{+} = \tilde{\mathcal{F}}_{\dot\alpha\beta} \;. \end{aligned} \tag{2.62}$$

Application of these relations to the general solution of the Bianchi identities of the chiral and anti-chiral representations (see eqs (2.30)),

$$\mathcal{F}_{a\alpha} = i\,(\sigma_a)_{\alpha\dot\beta}\,\overline{\mathcal{W}}_1^{\dot\beta} \;, \qquad \mathcal{F}_{a\dot\alpha} = i\,\mathcal{W}_2^{\beta}\,(\sigma_a)_{\beta\dot\alpha}$$

resp.

$$\tilde{\mathcal{F}}_{a\alpha} = i\,(\sigma_a)_{\alpha\dot\beta}\,\overline{\tilde{\mathcal{W}}}_1^{\dot\beta} \;, \qquad \tilde{\mathcal{F}}_{a\dot\alpha} = i\,\tilde{\mathcal{W}}_2^{\beta}\,(\sigma_a)_{\beta\dot\alpha}$$

leads to the equalities*)

$$\tilde{\mathcal{W}}_1 = \mathcal{W}_2 \;, \qquad \tilde{\mathcal{W}}_2 = \mathcal{W}_1 \;. \tag{2.63a}$$

On the other hand, as a consequence of (2.59) these superfields are related by

$$\tilde{\mathcal{W}}_1 = e^{V}\,\mathcal{W}_1\,e^{-V} \;, \qquad \tilde{\mathcal{W}}_2 = e^{V}\,\mathcal{W}_2\,e^{-V} \;. \tag{2.63b}$$

Thus, by combining these equations we obtain the following <u>reality property of $\mathcal{W}_2^{\beta}$</u> :

$$\begin{aligned} (\mathcal{W}_2^{\beta})^{+} = \left(e^{-V}\,\tilde{\mathcal{W}}_2^{\beta}\,e^{V}\right)^{+} &= e^{V}\,\overline{\tilde{\mathcal{W}}}_2^{\dot\beta}\,e^{-V} \\ &= e^{V}\,\overline{\mathcal{W}}_1^{\dot\beta}\,e^{-V} \;. \end{aligned} \tag{2.64}$$

"$\overline{\mathcal{W}}_1^{\dot\beta}$ is the chiral representation Hermitean conjugate of $\mathcal{W}_2^{\beta}$".

---

*) By definition we have $(\mathcal{W}^{\alpha})^{+} = \overline{\mathcal{W}}^{\dot\alpha}$.

Summarizing these results we can say that the reality condition $\mathbb{A}^{\dagger} = -\mathbb{A}$ implies that the superfields $\mathcal{W}_2^{\beta}$, $\bar{\mathcal{W}}_1^{\dot{\beta}}$ occuring in the solution of the chiral representation Bianchi identities are not independent of each other, but related by the formal gauge transformation e^V (similarly for the anti-chiral basis fields $\tilde{\mathcal{W}}_2^{\beta}$ $\bar{\tilde{\mathcal{W}}}_1^{\dot{\beta}}$).

(iii) Explicit solution of the constraints

The explicit solution of the constraints $\mathcal{F}_{\underline{\alpha\beta}} = 0$ and $\tilde{\mathcal{F}}_{\underline{\alpha\beta}} = 0$ follows immediately from the definition (2.55) and the corresponding solution in the real representation, i.e. equations (2.32),(2.33),(2.48):

$$\varphi_{\alpha} = -\mathcal{W}^{-1} D_{\alpha} \mathcal{W} \quad , \quad \varphi_{\dot{\alpha}} = 0 \quad , \quad \varphi_{a} = -\frac{i}{4} \bar{D}_{\dot{\alpha}} \bar{\sigma}_{a}^{\dot{\alpha}\beta} \varphi_{\beta}$$

$$\tilde{\varphi}_{\alpha} = 0 \quad , \quad \tilde{\varphi}_{\dot{\alpha}} = -\mathcal{W} \bar{D}_{\dot{\alpha}} \mathcal{W}^{-1}, \quad \tilde{\varphi}_{a} = -\frac{i}{4} D^{\alpha} (\sigma_{a})_{\alpha\dot{\beta}} \tilde{\varphi}^{\dot{\beta}} \; . \tag{2.65}$$

Since $\mathcal{W} = e^V$ (with $V = V^{\dagger}$) all quantities depend only on the real prepotential V which transforms under the chiral/anti-chiral gauge group (see eq.(2.50)).

Explicit expressions for some components of $\mathcal{F}$ and $\tilde{\mathcal{F}}$ can easily be obtained from their definition and the explicit expressions (2.65):

$$\mathcal{F}_{a\dot{\alpha}} = -\bar{D}_{\dot{\alpha}} \varphi_{a} = i\left(-\frac{1}{8} \bar{D}^2 \mathcal{W}^{-1} D^{\beta} \mathcal{W}\right) (\sigma_a)_{\beta\dot{\alpha}} \equiv i \, \mathcal{W}_2^{\beta} (\sigma_a)_{\beta\dot{\alpha}}$$

$$\tilde{\mathcal{F}}_{a\alpha} = -D_{\alpha} \tilde{\varphi}_{a} = i (\sigma_a)_{\alpha\dot{\beta}} \left(\frac{1}{8} D^2 \mathcal{W} \bar{D}^{\dot{\beta}} \mathcal{W}^{-1}\right) \equiv i (\sigma_a)_{\alpha\dot{\beta}} \bar{\tilde{\mathcal{W}}}_1^{\dot{\beta}} \; . \tag{2.66}$$

Substituting $\mathcal{W} = e^V$ in these equations we clearly recognize the well-known solutions (2.3) of the Bianchi identities in the (anti-) chiral basis:

$$\mathcal{W}_2^{\alpha} = -\frac{1}{8} \bar{D}^2 e^{-V} D^{\alpha} e^{V} \tag{2.67a}$$

$$\bar{\tilde{\mathcal{W}}}_1^{\dot{\alpha}} = \frac{1}{8} D^2 e^{V} \bar{D}^{\dot{\alpha}} e^{-V} \; . \tag{2.67b}$$

(By taking the hermitean conjugate of the so-defined superfield $\bar{\tilde{\mathcal{W}}}_1^{\dot{\alpha}}$ we obtain

$$\tilde{\mathcal{W}}_1^{\alpha} \equiv \left(\bar{\tilde{\mathcal{W}}}_1^{\dot{\alpha}}\right)^{\dagger} = \mathcal{W}_2^{\alpha}$$

in accord with (2.63a)).

To summarize : In the chiral representation we have $\bar{\mathcal{D}}_{\dot{\alpha}}^{(\varphi)} = \bar{D}_{\dot{\alpha}}$ and the general solution of the Bianchi identities is given by the $\mathcal{G}$-valued chiral superfield (2.67a) and by the related superfield $\bar{\mathcal{W}}_1^{\dot{\beta}} = e^{-V} (\mathcal{W}_2^{\beta})^{\dagger} e^{V}$. All quantities can be expressed in terms of a single field, namely the real $\mathcal{G}$-valued superfield V. (Similarly for the anti-chiral basis.)

Finally we note that the gauge field Lagrangian of the real representation (eq.(2.42)) does not change its form upon passage to the (anti-) chiral representation ($\mathcal{W}_2^\alpha = \mathcal{U}\, W^\alpha\, \mathcal{U}^{-1}$):

$$S_W = \int d^4x \int d^2\theta \; \mathrm{Tr}\left(W^\alpha W_\alpha\right) + h.c. = \int d^4x \int d^2\theta \; \mathrm{Tr}\left(\mathcal{W}_2^\alpha \, \mathcal{W}_{2\alpha}\right) + h.c.$$

(iv) Matter fields

Under the formal gauge transformation $\mathcal{U}^{-1}$ which relates the real to the chiral basis, the covariantly chiral matter superfield Φ becomes a chiral one, $\phi = \mathcal{U}\Phi$:

$$\begin{aligned} 0 &= \bar{\mathcal{D}}^{(A)}_{\dot\alpha} \Phi = \left(\bar{D}_{\dot\alpha} - A_{\dot\alpha}\right)\left(\mathcal{U}^{-1}\phi\right) \\ &= \bar{D}_{\dot\alpha}\left(\mathcal{U}^{-1}\phi\right) + \mathcal{U}^{-1}\left(\bar{D}_{\dot\alpha}\mathcal{U}\right)\left(\mathcal{U}^{-1}\phi\right) \quad \left(\text{recall } A_{\dot\alpha} = -\,\mathcal{U}^{-1}\bar{D}_{\dot\alpha}\mathcal{U}\right) \\ &= \left(\bar{D}_{\dot\alpha}\mathcal{U}^{-1}\right)\phi + \mathcal{U}^{-1}\left(\bar{D}_{\dot\alpha}\phi\right) - \left(\bar{D}_{\dot\alpha}\mathcal{U}^{-1}\right)\phi \\ &= \mathcal{U}^{-1}\left(\bar{D}_{\dot\alpha}\phi\right) . \end{aligned}$$

Similarly under the transformation $\mathcal{V}^{-1}$ relating the real representation to the anti-chiral one the covariantly anti-chiral superfield $\Phi^\dagger$ becomes an anti-chiral one, $\phi^\dagger = \Phi^\dagger\, \mathcal{V}^{-1}$.

Accordingly the Lagrangian (2.47) goes over into the familiar interaction Lagrangian of the chiral/anti-chiral representation (eq.(2.2)):

$$\int d^4\theta \; \Phi^\dagger \Phi = \int d^4\theta \; \left(\phi^\dagger \mathcal{V}\right)\left(\mathcal{U}^{-1}\phi\right) = \int d^4\theta \; \phi^\dagger \mathcal{W} \phi = \int d^4\theta \; \phi^\dagger e^V \phi .$$

(v) Projection to space-time

Since "all" quantities of the chiral basis are expressible in terms of the real superfield V, it seems indicated to follow the original approach and to define the gauge field content in terms of this basic superfield. However it is instructive to study what happens if one considers the projection of the super-connection $\mathcal{A}$.

If we define $V| \equiv C$ ($C^\dagger = C$), the lowest spatial component of eq.(2.61a) is given by

$$- \varphi_a^\dagger \Big| = e^{C} \varphi_a\Big| \, e^{-C} - e^{C} \partial_a e^{-C} \quad . \tag{2.68}$$

Thus the superfield φ_a does not directly project to a real gauge potential v_a on space-time. Similarly the chiral supergauge parameter Λ does not project to a real x-space parameter, but only to a complex one, $\Lambda| \equiv \frac{1}{2}\sigma$.

However in the abelian case (2.68) becomes

$$\mathrm{Im}\, v_a = -\partial_a C \qquad \left(\text{where} \quad \varphi_a\Big| \equiv -\frac{i}{2} v_a \right)$$

and by (2.56) the space-time field $\mathrm{Re}\, v_a$ transforms with a real gauge parameter:

$$\mathrm{Re}\, v_a' = \mathrm{Re}\, v_a + \partial_a (\mathrm{Re}\, \sigma) \quad .$$

Therefore in this case the weak reality condition of the chiral basis is strong enough to guarantee that $\mathrm{Im}\, v_a$ is a pure gauge and that the physically relevant information about the gauge potential is contained in $\mathrm{Re}\, v_a$.

Nevertheless in the non-abelian case this is not true anymore and if we ensure the reality of v_a by requiring C to vanish, we break supersymmetry*) and automatically enforce the WZ-gauge for V.

(vi) Starting from the (anti-) chiral representation

For greater clarity we started from the real representation and introduced the other ones by certain formal gauge transformations. Obviously one can also proceed the other way round as we will now briefly illustrate for the case of the chiral basis.

Thus we start with a superconnection φ transforming under the chiral gauge group,

$$\varphi' = {}^{\Sigma_1}\varphi \quad , \quad \Sigma_1 = e^{i\Lambda} \quad , \quad \bar{D}_{\dot\alpha} \Lambda = 0$$

and satisfying the following constraints:

$$\mathcal{F}_{\underline{\alpha\beta}} = 0 \quad , \quad -\varphi_a^\dagger \cong \varphi_a \quad , \quad -\varphi_\alpha^\dagger \cong \varphi_{\dot\alpha} \quad . \tag{2.69a-c}$$

---

*) Under a susy transformation with parameter ξ we have:

$$\delta_\xi C = i\,\xi\chi - i\,\bar\xi\bar\chi$$

$$\delta_\xi \chi_\alpha = i\,(\sigma^m \bar\xi)_\alpha\, v_m + \text{terms in } C, M, N .$$

As a particular solution of $\mathcal{F}_{\alpha\beta} = 0 = \mathcal{F}_{\dot{\alpha}\dot{\beta}}$ you have

$$\varphi_\alpha = - \mathcal{W}^{-1} D_\alpha \mathcal{W} \quad , \quad \varphi_{\dot{\alpha}} = 0 \tag{2.70}$$

where $\mathcal{W} = e^V$ with V being an <u>arbitrary</u> $\mathcal{G}$-valued superfield. Consequently

$$-\varphi_\alpha^\dagger = - e^{V^\dagger} \bar{D}_{\dot{\alpha}} e^{-V^\dagger}$$

which is a pure gauge and thereby formally gauge equivalent to $0 = \varphi_{\dot{\alpha}}$:

$$-\varphi_\alpha^\dagger = e^{V^\dagger} \varphi_{\dot{\alpha}} e^{-V^\dagger} - e^{V^\dagger} \bar{D}_{\dot{\alpha}} e^{-V^\dagger} .$$

In view of this equation the reality condition (2.69b) is to be specified as follows:

$$-\varphi_a^\dagger = e^{V^\dagger} \varphi_a e^{-V^\dagger} - e^{V^\dagger} \partial_a e^{-V^\dagger} . \tag{2.71}$$

Since $\mathcal{F}_{\alpha\dot{\beta}} = 0$ the relations (2.70) imply

$$\varphi_a = \frac{i}{4} \bar{D}_{\dot{\alpha}} \bar{\sigma}_a^{\dot{\alpha}\beta} \mathcal{W}^{-1} D_\beta \mathcal{W} \qquad (\mathcal{W} = e^V)$$

and insertion of this expression into (2.71) shows after a short calculation that this condition can only be satisfied if $V^\dagger = V$.

Thus we have recovered the previously presented description of the chiral representation.

II.6 Summary

In the following we schematically summarize the previously obtained results. We always assume that

$$\mathcal{U}^\dagger = \mathcal{V}^{-1} \quad , \quad \mathcal{W} = \mathcal{V} \cdot \mathcal{U}^{-1} = e^V \quad (V^\dagger = V)$$

$$\Pi^\dagger = \Sigma^{-1} \quad , \quad \Sigma = e^{i\Lambda} \quad (\bar{D}_{\dot{\alpha}} \Lambda = 0) .$$

TABLE 1

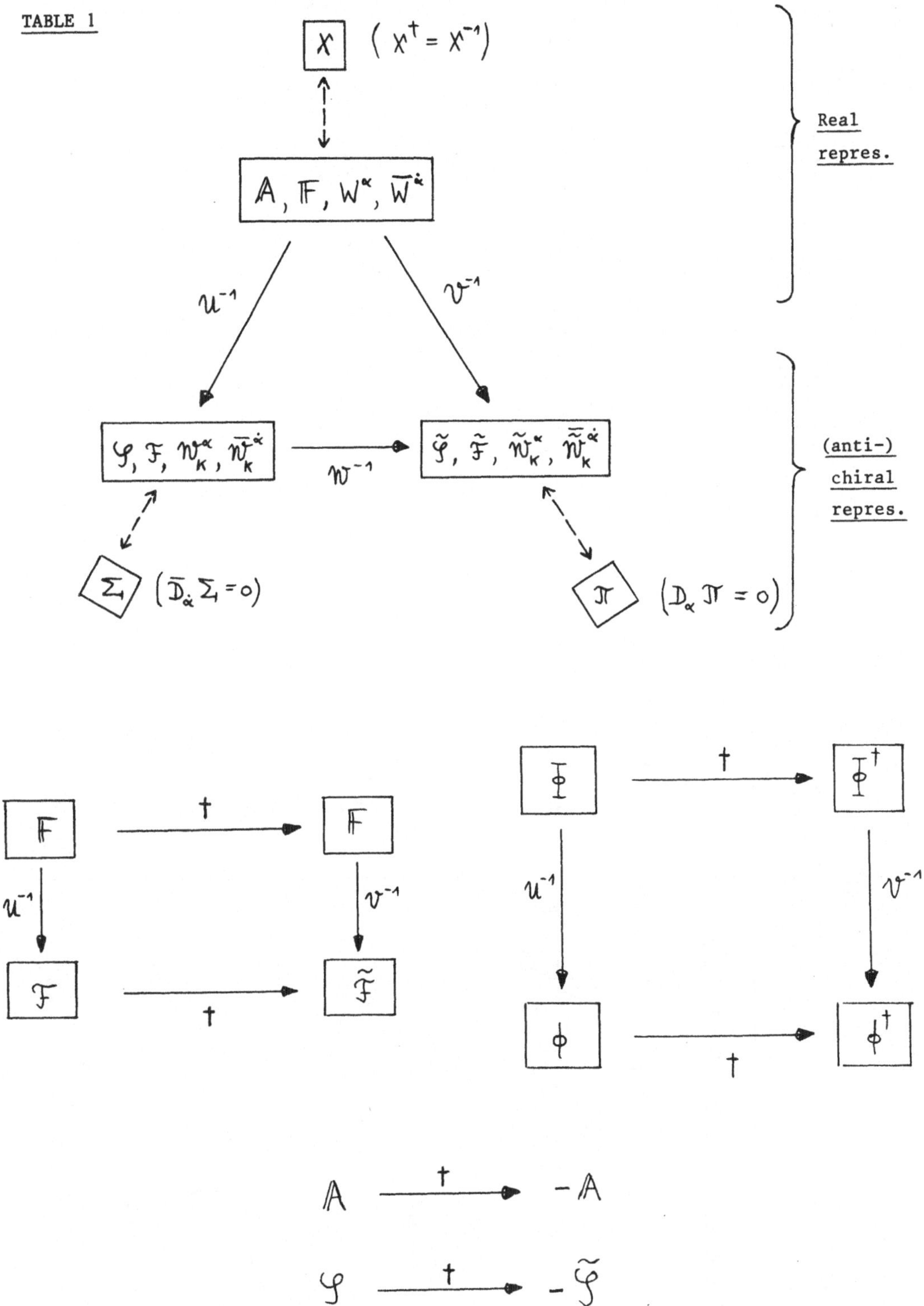

$(X^\dagger = X^{-1})$
Real
repres.
(anti-)
chiral
repres.
$(\bar{D}_{\dot\alpha}\Sigma = 0)$
$(D_\alpha \Pi = 0)$

TABLE 2

$\longleftarrow$ pass to the right side $\quad \mathbb{A} \quad \xrightarrow[(\mathcal{U}' = \Sigma_1^{-1}\mathcal{U}X)]{\mathcal{U}^{-1}} \quad \varphi \quad \xrightarrow[(\mathcal{W}' = \Pi^{-1}\mathcal{W}\Sigma_1)]{\mathcal{W}^{-1}} \quad \tilde{\varphi} \quad \xleftarrow[(\mathcal{V}' = \Pi^{-1}\mathcal{V}X)]{\mathcal{V}^{-1}}$

$\mathbb{A}$	φ	$\tilde{\varphi}$
Connection- and curvature-superforms:		
$\mathbb{A} = E^A \mathbb{A}_A = -\mathbb{A}^\dagger$	$\varphi = E^A \varphi_A$	$\tilde{\varphi} = E^A \tilde{\varphi}_A$
$\mathbb{F} = d\mathbb{A} + \mathbb{A}\mathbb{A} = \frac{1}{2} E^A E^B \mathbb{F}_{BA} = \mathbb{F}^\dagger$	$\mathcal{F} = d\varphi + \varphi\varphi = \frac{1}{2} E^A E^B \mathcal{F}_{BA}$	$\tilde{\mathcal{F}} = d\tilde{\varphi} + \tilde{\varphi}\tilde{\varphi} = \frac{1}{2} E^A E^B \tilde{\mathcal{F}}_{BA} = \mathcal{F}^\dagger$
Supergauge group:		
$X = e^{\mathfrak{X}}, \quad \mathfrak{X}^\dagger = -\mathfrak{X}$	$\Sigma_1 = e^{i\Lambda}, \quad \bar{D}_{\dot\alpha}\Lambda = 0$	$\Pi = e^{i\Lambda^\dagger}, \quad D_\alpha \Lambda^\dagger = 0$
Bianchi identities:		
$\mathcal{D}^{(\mathbb{A})}\mathbb{F} = 0$	$\mathcal{D}^{(\varphi)}\mathcal{F} = 0$	$\mathcal{D}^{(\tilde{\varphi})}\tilde{\mathcal{F}} = 0$
Constraints:		
$0 = \mathbb{F}_{\alpha\beta} = \mathbb{F}_{\dot\alpha\dot\beta} = \mathbb{F}_{\alpha\dot\beta}$	$0 = \mathcal{F}_{\alpha\beta} = \mathcal{F}_{\dot\alpha\dot\beta} = \mathcal{F}_{\alpha\dot\beta}$	$0 = \tilde{\mathcal{F}}_{\alpha\beta} = \tilde{\mathcal{F}}_{\dot\alpha\dot\beta} = \tilde{\mathcal{F}}_{\alpha\dot\beta}$

Reality condition:

$-\mathbb{A}^{\dagger} = \mathbb{A}$	$-\varphi^{\dagger} = \mathcal{W}\varphi\mathcal{W}^{-1} - \mathcal{W}d\mathcal{W}^{-1}$	$-\tilde{\varphi}^{\dagger} = \mathcal{W}^{-1}\tilde{\varphi}\mathcal{W} - \mathcal{W}^{-1}d\mathcal{W}$

Solution of the constraints:

$\mathbb{F}_{\dot\alpha\dot\beta} = 0: \quad \mathbb{A}_{\dot\alpha} = -\mathcal{U}^{-1}\bar{D}_{\dot\alpha}\mathcal{U}$	$\varphi_{\dot\alpha} = 0$	$\tilde{\varphi}_{\dot\alpha} = -\mathcal{W}\bar{D}_{\dot\alpha}\mathcal{W}^{-1} = -\varphi_{\alpha}^{\dagger}$
$\mathbb{F}_{\alpha\beta} = 0: \quad \mathbb{A}_{\alpha} = -\mathcal{V}^{-1}D_{\alpha}\mathcal{V} = -\mathbb{A}_{\dot\alpha}^{\dagger}$	$\varphi_{\alpha} = -\mathcal{W}^{-1}D_{\alpha}\mathcal{W}$	$\tilde{\varphi}_{\alpha} = 0$
$\mathbb{F}_{\alpha\dot\beta} = 0: \mathbb{A}_{a} = \frac{-i}{4}\left(D^{\alpha}\sigma^{a}_{\alpha\dot\beta}\mathbb{A}^{\dot\beta} + \bar{D}_{\dot\alpha}(\bar\sigma^{a})^{\dot\alpha\beta}\mathbb{A}_{\beta} - [\mathbb{A}^{\alpha}, \sigma^{a}_{\alpha\dot\beta}\mathbb{A}^{\dot\beta}]\right)$	$\varphi_{a} = -\frac{i}{4}\bar{D}_{\dot\alpha}(\bar\sigma_{a})^{\dot\alpha\beta}\varphi_{\beta}$	$\tilde{\varphi}_{a} = -\frac{i}{4}D^{\alpha}(\sigma_{a})_{\alpha\dot\beta}\tilde{\varphi}^{\dot\beta} = -\varphi_{a}^{\dagger}$

In the supergauge $\mathcal{U} = e^{-V/2} = \mathcal{U}^{\dagger}$, $\mathcal{V} = e^{V/2} = \mathcal{V}^{\dagger}$	Writing $\mathcal{W} = e^{V}$:	
$\mathbb{A}_{\dot\alpha} = -e^{\frac{1}{2}V}\bar{D}_{\dot\alpha}e^{-\frac{1}{2}V}$	$\varphi_{\dot\alpha} = 0$	$\tilde{\varphi}_{\dot\alpha} = -e^{V}\bar{D}_{\dot\alpha}e^{-V}$
$\mathbb{A}_{\alpha} = -e^{-\frac{1}{2}V}D_{\alpha}e^{\frac{1}{2}V}$	$\varphi_{\alpha} = -e^{-V}D_{\alpha}e^{V}$	$\tilde{\varphi}_{\alpha} = 0$

In the WZ-gauge $(V^3 = 0)$: $(\mathbf{1} = \mathcal{U}| = \mathcal{V}| = \mathcal{W}|)$

$\mathbb{A}_{\alpha}\| = 0 = \mathbb{A}_{\dot\alpha}\|$	$\varphi_{\alpha}\| = 0 = \varphi_{\dot\alpha}\|$	$\tilde{\varphi}_{\alpha}\| = 0 = \tilde{\varphi}_{\dot\alpha}\|$

Solution of the Bianchi-id. subject to the constraints:		
$\mathbb{F}_{a\alpha} = i\,(\sigma_a)_{\alpha\dot\beta}\,\overline{W}^{\dot\beta}$	$\mathcal{F}_{a\alpha} = i\,(\sigma_a)_{\alpha\dot\beta}\,\overline{\mathcal{W}}_1^{\dot\beta}$	$\tilde{\mathcal{F}}_{a\alpha} = i\,(\sigma_a)_{\alpha\dot\beta}\,\overline{\tilde{\mathcal{W}}}_1^{\dot\beta} = -\,\mathcal{F}_{a\dot\alpha}^{\dagger}$
$\mathbb{F}_{a\dot\alpha} = i\,W^{\beta}(\sigma_a)_{\beta\dot\alpha} = -\,\mathbb{F}_{a\alpha}^{\dagger}$	$\mathcal{F}_{a\dot\alpha} = i\,\mathcal{W}_2^{\beta}\,(\sigma_a)_{\beta\dot\alpha}$	$\tilde{\mathcal{F}}_{a\dot\alpha} = i\,\tilde{\mathcal{W}}_2^{\beta}\,(\sigma_a)_{\beta\dot\alpha}$
$\mathbb{F}_{ab} = \frac{1}{2}\left(\mathcal{D}^{(A)}\sigma_{ab}W - \overline{\mathcal{D}}^{(A)}\bar\sigma_{ab}\overline{W}\right) = -\,\mathbb{F}_{ab}^{\dagger}$	$\mathcal{F}_{ab} = \frac{1}{2}\left(\mathcal{D}^{(\varphi)}\sigma_{ab}\mathcal{W}_2 - \overline{\mathcal{D}}^{(\varphi)}\bar\sigma_{ab}\overline{\mathcal{W}}_1\right)$	$\tilde{\mathcal{F}}_{ab} = \frac{1}{2}\left(\mathcal{D}^{(\tilde\varphi)}\sigma_{ab}\tilde{\mathcal{W}}_2 - \overline{\mathcal{D}}^{(\tilde\varphi)}\bar\sigma_{ab}\overline{\tilde{\mathcal{W}}}_1\right) = -\,\mathcal{F}_{ab}^{\dagger}$
with:		
$\overline{\mathcal{D}}_{\dot\alpha}^{(A)}W_{\beta} = 0 = \mathcal{D}_{\alpha}^{(A)}\overline{W}_{\dot\beta}$	$\overline{\mathcal{D}}_{\dot\alpha}^{(\varphi)}\mathcal{W}_2^{\beta} = 0 = \mathcal{D}_{\alpha}^{(\varphi)}\overline{\mathcal{W}}_1^{\dot\beta}$	$\overline{\mathcal{D}}_{\dot\alpha}^{(\tilde\varphi)}\tilde{\mathcal{W}}_2^{\beta} = 0 = \mathcal{D}_{\alpha}^{(\tilde\varphi)}\overline{\tilde{\mathcal{W}}}_1^{\dot\beta}$
$\mathcal{D}^{(A)\alpha}W_{\alpha} = \overline{\mathcal{D}}_{\dot\alpha}^{(A)}\overline{W}^{\dot\alpha}$	$\mathcal{D}^{(\varphi)\alpha}\mathcal{W}_{2\alpha} = \overline{\mathcal{D}}_{\dot\alpha}^{(\varphi)}\overline{\mathcal{W}}_1^{\dot\alpha}$	$\mathcal{D}^{(\tilde\varphi)\alpha}\tilde{\mathcal{W}}_{2\alpha} = \overline{\mathcal{D}}_{\dot\alpha}^{(\tilde\varphi)}\overline{\tilde{\mathcal{W}}}_1^{\dot\alpha}$
and transformation properties		
$W'_{\underline{\alpha}} = X^{-1}\,W_{\underline{\alpha}}\,X \quad (X^{\dagger} = X^{-1})$	$\mathcal{W}' = \Sigma_1^{-1}\,\mathcal{W}\,\Sigma_1 \quad (\bar{D}_{\dot\alpha}\Sigma_1 = 0)$	$\tilde{\mathcal{W}}' = \Pi^{-1}\,\tilde{\mathcal{W}}\,\Pi \quad (D_{\alpha}\Pi = 0)$
Explicit expressions:		
	$\mathcal{W}_2^{\alpha} = -\frac{1}{8}\bar{D}^2\mathcal{W}^{-1}D^{\alpha}\mathcal{W} \quad \left(= \frac{1}{4}\bar{D}^2\varphi^{\alpha}\right)$	$\tilde{\mathcal{W}}_1^{\alpha} = \mathcal{W}_2^{\alpha}$
	$\overline{\mathcal{W}}_2^{\dot\alpha} = \overline{\tilde{\mathcal{W}}}_1^{\dot\alpha}$	$\overline{\tilde{\mathcal{W}}}_1^{\dot\alpha} = \frac{1}{8}D^2\mathcal{W}\bar{D}^{\dot\alpha}\mathcal{W}^{-1} \quad \left(= \frac{1}{4}D^2\tilde\varphi^{\dot\alpha}\right)$

<u>Lagrangian for the gauge field:</u>

$\int d^2\theta \; \mathrm{Tr}\, W^{\alpha} W_{\alpha} \quad + \quad h.c.$	$\int d^2\theta \; \mathrm{Tr}\, \mathcal{W}_2^{\alpha} \mathcal{W}_{2\alpha} \quad + \quad h.c.$

<u>Component fields:</u>

$W_{\alpha}\vert \equiv -\frac{i}{2}\lambda_{\alpha}$	$\mathcal{W}_{2\alpha}\vert \equiv -\frac{i}{2}\lambda_{\alpha}$
$\mathcal{D}^{(A)\alpha} W_{\alpha}\vert \equiv -\mathbb{D}$	$D^{\alpha}\mathcal{W}_{2\alpha}\vert \equiv -D$
$\mathcal{D}^{(A)2} W_{\alpha}\vert = -2\,\mathcal{D}_{\alpha\dot{\alpha}}\bar{\lambda}^{\dot{\alpha}} \quad \left(A_m\vert \equiv \frac{-i}{2} v_m\right)$	$D^2\mathcal{W}_{2\alpha}\vert \equiv -2\,\mathcal{D}_{\alpha\dot{\alpha}}\bar{\lambda}^{\dot{\alpha}}$

<u>Relation between the components in the different repres. in the WZ-gauge:</u>

$$\lambda_{\alpha} = \lambda_{\alpha} \quad , \quad -\mathbb{D} = D^{\alpha}W_{\alpha}\vert = D^{\alpha}\mathcal{W}_2^{\alpha}\vert = -D \quad , \quad D^2 W_{\alpha}\vert = D^2\mathcal{W}_{2\alpha}\vert \quad (v_m = v_m)$$

TABLE 3

$\xleftarrow{\text{pass to the right side}} \Phi^\dagger \xleftarrow{\;\dagger\;} \Phi \xrightarrow[(\mathcal{U}' = \Sigma^{-1}\mathcal{U}X)]{\mathcal{U}^{-1}} \phi\,(=\mathcal{U}\Phi) \xrightarrow{\;\dagger\;} \phi^\dagger\,(=\Phi^\dagger\mathcal{V}^{-1}) \xleftarrow[(\mathcal{V}' = \Pi^{-1}\mathcal{V}X)]{\mathcal{V}^{-1}}$

Character. of the superfield:			
$\mathcal{D}^{(A)}_\alpha \Phi^\dagger = 0$	$\bar{\mathcal{D}}^{(A)}_{\dot\alpha} \Phi = 0$	$\bar{D}_{\dot\alpha} \phi = 0$	$D_\alpha \phi^\dagger = 0$
Component-fields:			
$\Phi^\dagger\vert = A^\dagger$	$\Phi\vert \equiv A$	$\phi\vert \equiv A$	$\phi^\dagger\vert = A^\dagger$
$\bar{\mathcal{D}}^{(A)}_{\dot\alpha} \Phi^\dagger\vert = \sqrt{2}\;\bar{\Psi}_{\dot\alpha}$	$\mathcal{D}^{(A)}_\alpha \Phi\vert \equiv \Psi_\alpha$	$D_\alpha \phi\vert \equiv \sqrt{2}\;\psi_\alpha$	$\bar{D}_{\dot\alpha} \phi^\dagger\vert = \sqrt{2}\;\bar{\psi}_{\dot\alpha}$
$\bar{\mathcal{D}}^{(A)2} \Phi^\dagger\vert = -4\,F^\dagger$	$\mathcal{D}^{(A)2} \Phi\vert \equiv -4\,F$	$D^2 \phi\vert \equiv -4\,F$	$\bar{D}^2 \phi^\dagger\vert = -4\,F^\dagger$
Gauge transform. of the superfields:			
$\Phi^{\dagger\prime} = {}^{X}\Phi^\dagger \equiv \Phi^\dagger X$	$\Phi' = {}^{X}\Phi \equiv X^{-1}\Phi$	$\phi' = {}^{\Sigma}\phi \equiv \Sigma^{-1}\phi$	$\phi^{\dagger\prime} = {}^{\Pi}\phi^\dagger \equiv \phi^\dagger \Pi$

Interaction-Lagrangian:

$$\int d^4x \int d^2\theta \; \underline{\Phi}^\dagger \underline{\Phi} \qquad \Big| \qquad \int d^4x \int d^4\theta \; \phi^\dagger \, \mathcal{V} \, \mathcal{U}^{-1} \phi = \int d^4x \int d^4\theta \; \phi^\dagger e^V \phi$$

Relations between the components in the different representations

$$\underline{A} = \mathcal{U}^{-1}| \cdot A$$

$$\sqrt{2}\, \underline{\Psi}_\alpha = \mathcal{U}^{-1}| \cdot \left[\sqrt{2}\, \Psi_\alpha + (\mathcal{W}^{-1} D_\alpha \mathcal{W})| \cdot A\right]$$

$$-4\, \underline{F} = \mathcal{U}^{-1}| \cdot \left[-4F + (\mathcal{W}^{-1} D^2 \mathcal{W})| \cdot A + 2\sqrt{2}\, (\mathcal{W}^{-1} D^\alpha \mathcal{W})| \cdot \Psi_\alpha\right]$$

Relations in the WZ-gauge $(\mathbb{1} = \mathcal{W}| = \mathcal{U}| = \mathcal{V}|)$:

$$\underline{A} = A \quad , \quad \underline{\Psi}_\alpha = \Psi_\alpha \quad , \quad \underline{F} = F$$

II.7 What's the use of all this?

So far we have shown that the original formulation of SYM-theories can be recovered as a special case from a more systematic and geometric approach. A natural question to ask is whether this approach and in particular the real representation have a further reaching significance and usefulness.

As will be shown in the following parts the real basis is well suited for describing the classical theory: by gauge covariant projection one introduces only the minimal number of space-time fields and easily determines the corresponding supersymmetry and BRS transformations as well as the action. A serious shortcoming of this approach is that the constrained YM-covariant derivatives represent the susy algebra only up to field dependent gauge transformations. Thus one is automatically in a WZ-gauge like situation. For the chiral basis where the gauge field content is defined in terms of the real prepotential V this is not the case unless one explicitly imposes the WZ-supergauge on V.

Although there exist quantization schemes for constrained fields, it is definitely more convenient to rely on unconstrained fields which in our context, means unconstrained superfields*). Thus one is led to consider the quantization of the (unconstrained) prepotentials either in the real or in the chiral/anti-chiral basis. Obviously the latter is much more convenient for there is only one prepotential and one symmetry group whereas in the other case one has two (related) fields and two invariance groups (real gauge transformations and chiral pregauge transformations). For these reasons one generally quantizes the superfield V and uses the real representation fields only as background fields[57),114)].

The geometric set-up presented here can also be applied for the algebraic determination of chiral anomalies in SYM-theories : by using the explicit solution of the constraints in the chiral representation one can derive a superspace expression for the chiral anomaly in terms of the basic superfield V [26),62),97)] (section IV.2 (i)).

---

*)This does not apply to the chiral superfield, though in this case one always has the possibility for explicitly solving the constraint in terms of an unconstrained field, $\phi = \bar{D}^2 F$ [155)].

III. CLASSICAL SYM-THEORIES IN THE GAUGE REAL REPRESENTATION

The usefulness of the real basis for the quantization of SYM-theories is rather limited and therefore the standard references on the subject either do not discuss it at all or not to the same extent than the chiral representation. Yet the real basis provides a very simple and direct description of the classical theory which generalizes in a straightforward way to extended supersymmetry and to supergravity ; for these reasons we will present it in some detail in the following.

III.1 Component fields

In the case at hand the general solution of the Bianchi identities is given by a single Lie algebra valued and covariantly chiral superfield W_α (see section II.4 (i)). The independent components of W_α can be used to specify the field content of the gauge multiplet. They are most conveniently defined by the projection method outlined in section I.11 (ii) ; but, since W_α is covariantly chiral, we project with the gauge covariant derivatives $\mathcal{D}_\alpha, \bar{\mathcal{D}}_{\dot\alpha}$ instead of the left-invariant ones $(D_\alpha, \bar{D}_{\dot\alpha})$. The independent component fields of W_α are then given by

$$W_\alpha\Big| \equiv \frac{-i}{2}\lambda_\alpha \quad , \quad \mathcal{D}^\alpha W_\alpha\Big| \equiv -\mathbb{D} \quad , \quad \mathcal{D}^\beta\mathcal{D}_\beta W_\alpha\Big| \equiv -2\,\mathcal{D}_{\alpha\dot\alpha}\bar{\lambda}^{\dot\alpha} \tag{3.1}$$

where $\mathbb{D}$ is real, $\mathbb{D}^\dagger = \mathbb{D}$. (We denote the x-fields with a double bar to distinguish them from the differently defined fields of the chiral representation; all supergauge covariant derivatives of part III are to be taken with respect to the superconnection $\mathbb{A} = -\mathbb{A}^\dagger$.) The component which is indirectly introduced by the last equality is the YM-potential $\mathbb{v}_a$ occuring in the ordinary covariant derivative

$$\mathcal{D}_{\alpha\dot\alpha}\bar{\lambda}^{\dot\alpha} \equiv \sigma^a_{\alpha\dot\alpha}\left(\mathcal{D}_a\bar{\lambda}^{\dot\alpha}\right) = \sigma^a_{\alpha\dot\alpha}\left(\partial_a\bar{\lambda}^{\dot\alpha} + \frac{i}{2}\left[\mathbb{v}_a\,,\bar{\lambda}^{\dot\alpha}\right]\right). \tag{3.2}$$

We remark that this definition of $\mathbb{v}_a$ is consistent with our previous one, eq. (2.44a). Indeed, as a consequence of the constraint $\mathbb{F}_{\alpha\dot\alpha} = 0$ we have

$$\left\{\mathcal{D}_\alpha\,,\bar{\mathcal{D}}_{\dot\alpha}\right\} = -2i\,\sigma^a_{\alpha\dot\alpha}\,\mathcal{D}_a$$

and application to the covariantly anti-chiral superfield $\bar{W}^{\dot\alpha}$ leads to

$$-2i\,\sigma^a_{\alpha\dot\alpha}\,\mathcal{D}_a\bar{W}^{\dot\alpha} = \mathcal{D}_\alpha\left(\bar{\mathcal{D}}_{\dot\alpha}\bar{W}^{\dot\alpha}\right).$$

Now $\bar{\mathcal{D}}_{\dot\alpha}\bar{W}^{\dot\alpha} = \mathcal{D}^\alpha W_\alpha$ (eq.(2.30)) and $\{\mathcal{D}_\alpha, \mathcal{D}_\beta\} = -\mathbb{F}_{\alpha\beta} = 0$, therefore

$$- 2i\, \sigma^{a}_{\alpha\dot\alpha} \left(\mathcal{D}_a \overline{W}^{\dot\alpha} \right) = \mathcal{D}_\alpha \left(\mathcal{D}^\beta W_\beta \right) = - \mathcal{D}_\alpha \mathcal{D}_\beta W^\beta = - \left(\tfrac{1}{2} \varepsilon_{\alpha\beta} \mathcal{D}^\gamma \mathcal{D}_\gamma \right) W^\beta$$
$$= - \tfrac{1}{2} \mathcal{D}^\gamma \mathcal{D}_\gamma W_\alpha \quad .$$

Finally, comparison of the lowest component of this expression,

$$\mathcal{D}^\beta \mathcal{D}_\beta W_\alpha = 4i\, \sigma^{a}_{\alpha\dot\alpha} \left(\mathcal{D}_a \overline{W}^{\dot\alpha} \right)$$
$$= 4i\, \sigma^{a}_{\alpha\dot\alpha} \left(\partial_a \overline{W}^{\dot\alpha} - [\mathbb{A}_a , \overline{W}^{\dot\alpha}] \right)$$

with equation (3.2) yields our previous definition

$$\mathbb{A}_a \big| = - \frac{i}{2} \mathbb{V}_a \quad .$$

The components of the covariantly chiral matter field Φ are to be defined in a similar way:

$$\Phi \big| \equiv A \quad , \quad \mathcal{D}_\alpha \Phi \big| \equiv \sqrt{2}\, \Psi_\alpha \quad , \quad \mathcal{D}^\alpha \mathcal{D}_\alpha \Phi \big| \equiv -4F \tag{3.3}$$

Strictly speaking the derivatives involved in the equations (3.1),(3.3) are not only covariant with respect to the YM-connection $\mathbb{A}$, but also w.r.t. the canonical linear connection $\phi \equiv 0$; on flexible (curved) superspace the latter is simply "replaced" by the non-trivial Lorentz connection.

III.2 The "susy transformations" and their algebra

(i) Definition of the transformations

Since we used the gauge covariant derivatives to introduce the component fields, we also rely on them to define the susy transformations of the superfields:

$$\delta_\xi F := \mathcal{L}_\xi F = \xi^A \mathcal{D}_A F \tag{3.4}$$

where $\mathcal{L}_\xi \equiv i_\xi \mathcal{D} + \mathcal{D} i_\xi$ represents the YM-covariant Lie derivative in the direction of the susy generating vector field (1.5) (suppose $\xi^a| = a^a = 0$). These transformations correspond to the sum of an ordinary susy transformation and of a supergauge transformation with field dependent parameter $\xi^A \mathbb{A}_A$.

Application of (3.4) and of the rule $(\delta_\xi F|) \equiv (\delta_\xi F)|$ to the superfields $F = W_\alpha, \mathcal{D}^\alpha W_\alpha, \Phi, \mathcal{D}_\alpha \Phi, \mathcal{D}^2 \Phi$ straightforwardly leads to the familiar "susy transformations in the WZ-gauge", equations (2.10). For instance

$$\delta_\zeta A = \delta_\xi(\Phi|) = (\delta_\xi \Phi)| = (\xi^A \mathcal{D}_A \Phi)| = \zeta^\alpha (\mathcal{D}_\alpha \Phi)| = \sqrt{2}\, \zeta^\alpha \Psi_\alpha$$

$$\sqrt{2}\, \delta_\zeta \Psi_\alpha = (\xi^B \mathcal{D}_B \mathcal{D}_\alpha \Phi)| = \zeta^\beta \mathcal{D}_\beta \mathcal{D}_\alpha \Phi| + \bar{\zeta}_{\dot\beta}\, \bar{\mathcal{D}}^{\dot\beta} \mathcal{D}_\alpha \Phi|$$

$$= \frac{1}{2} \varepsilon_{\beta\alpha} \zeta^\beta (\mathcal{D}^\gamma \mathcal{D}_\gamma \Phi)| - \bar{\zeta}^{\dot\beta} \{\bar{\mathcal{D}}_{\dot\beta}, \mathcal{D}_\alpha\} \Phi|$$

$$= -\frac{1}{2} \zeta_\alpha (-4F) - \bar{\zeta}^{\dot\beta} (-2i\, \sigma^a_{\alpha\dot\beta})(\mathcal{D}_a \Phi)|$$

$$= 2\, \zeta_\alpha F + 2i\, (\sigma^a_{\alpha\dot\beta} \bar{\zeta}^{\dot\beta})(\mathcal{D}_a A)$$

hence

$$\delta_\zeta \Psi_\alpha = i\sqrt{2}\, (\sigma^m \bar{\zeta})_\alpha \mathcal{D}_m A + \sqrt{2}\, \zeta_\alpha F .$$

Since the connection does not transform tensorially under the gauge group, the transformation law (3.4) does not apply to $\mathbb{A}_A$; the appropriate generalization is given by

$$\delta_\xi \mathbb{A}_A = (i_\xi \mathbb{F})_A = \xi^B \mathbb{F}_{BA} \tag{3.5}$$

and leads to the familiar susy transformation law for the YM-potential $\mathbf{v}_a$.

(ii) Their algebra

As a consequence of the constraints the algebra of the "susy transformations" (3.4) is given by

$$\begin{aligned}[\delta_\zeta, \delta_\eta] F| &= [\zeta\mathcal{D} + \bar{\zeta}\bar{\mathcal{D}}, \eta\mathcal{D} + \bar{\eta}\bar{\mathcal{D}}] F| \\ &= -2i(\zeta\sigma^m\bar{\eta} - \eta\sigma^m\bar{\zeta})(\mathcal{D}_m F)| \\ &= -2i(\zeta\sigma^m\bar{\eta} - \eta\sigma^m\bar{\zeta})\left(\partial_m F| + \frac{i}{2}[\mathbf{v}_m, F|]\right).\end{aligned} \tag{3.6}$$

Thus we have the following situation. In the gauge real representation one only introduces the minimal number of space-time fields and one automatically has a WZ-gauge like situation where the susy algebra is only represented modulo field dependent gauge transformations. Somehow the gauge covariantized susy transformations (3.4) represent in a compact way the supersymmetry and compensating supergauge transformations of the original approach to the theory. This comparison can be made more precise by considering the YM-covariant derivatives of the anti-chiral basis : if we use the explicit solution (2.65) of the constraints in terms of the real superfield V and go to the WZ-gauge (2.6a), we have

$$(\zeta^\alpha \mathcal{D}_\alpha + \bar{\zeta}_{\dot\alpha} \bar{\mathcal{D}}^{\dot\alpha})\phi = (\zeta D + \bar\zeta \bar D)\phi - (\bar\zeta_{\dot\alpha} \tilde\varphi^{\dot\alpha})\phi \tag{3.7}$$

with

$$\begin{aligned} -(\bar\zeta_{\dot\alpha}\tilde\varphi^{\dot\alpha})\phi &= \bar\zeta_{\dot\alpha}(e^V \bar D^{\dot\alpha} e^{-V})\phi \\ &= \left[(\theta\sigma^m\bar\zeta)v_m - i(\theta\theta)(\bar\zeta\bar\lambda) + 2i(\bar\zeta\bar\theta)(\theta\lambda) + \left\{\begin{matrix}\text{higher}\\ \text{order}\\ \text{terms}\end{matrix}\right\}\right]\phi . \end{aligned}$$

The last expression corresponds exactly to the compensating supergauge transformation $\delta_g\phi = -i\Lambda\phi$ considered in (2.9). (Note that only the first two terms of the last equation contribute to the "susy transformations" (3.7) and to their algebra.)

III.3 The Lagrangian

(i) Massless matter and gauge fields

We consider the supergauge invariant interaction of the massless matter and gauge multiplets Φ and W_α and sketch the derivation of the component field expression for the corresponding Lagrangian. The latter is given by (2.42) and (2.47),

$$S = \int d^4x \int d^4\theta\, \Phi^\dagger \Phi + \int d^4x \left[\int d^2\theta\, \mathrm{Tr}(W^\alpha W_\alpha) + \text{h.c.}\right] \tag{3.8}$$

and can be evaluated by the standard technique of replacing the fermionic integrations by spinorial derivatives (eq.(1.93)); since the integrands are supergauge invariant these derivatives can actually be replaced by YM-covariant ones :

$$\begin{aligned} S &= \int d^4x \int d^4\theta\, \Phi^\dagger\Phi + \left\{\int d^4x \int d^2\theta\, \mathrm{Tr}(W^\alpha W_\alpha) + \text{h.c.}\right\} \\ &= \frac{1}{16}\int d^4x \left(\bar{\mathcal{D}}^2 \mathcal{D}^2 \Phi^\dagger \Phi\right)\Big| - \frac{1}{4}\left\{\int d^4x\, \mathrm{Tr}\left(\mathcal{D}^2 W^\alpha W_\alpha\right)\Big| + \text{h.c.}\right\} . \end{aligned}$$

We begin with an evaluation of the first term which we denote by S_ϕ. The relation $\mathcal{D}_\alpha \Phi^+ = 0$ implies

$$S_\Phi = \frac{1}{16} \int d^4x \; \bar{\mathcal{D}}^2 \left(\Phi^+ \mathcal{D}^2 \Phi \right) \Big|$$

$$= \frac{1}{16} \int d^4x \left\{ \left(\bar{\mathcal{D}}^2 \Phi^+ \right) \left(\mathcal{D}^2 \Phi \right) + 2 \left(\bar{\mathcal{D}}_{\dot\alpha} \Phi^+ \right) \left(\bar{\mathcal{D}}^{\dot\alpha} \mathcal{D}^2 \Phi \right) + \Phi^+ \left(\bar{\mathcal{D}}^2 \mathcal{D}^2 \Phi \right) \right\} \Big| .$$

The first contribution to this expression follows immediately from the definition (3.3):

$$\frac{1}{16} \int d^4x \, (-4 F^+)(-4F) = \int d^4x \; F^+ F \quad .$$

To determine the other contributions one has to use the commutation relations of the covariant derivatives, the constraints and the Bianchi identities in order to obtain expressions in which the components of Φ, Φ^+ resp. W_α, $\bar W_{\dot\alpha}$ occur explicitly.

The same method has to be applied to the two (identical) gauge field contributions:

$$S_W = \frac{-2}{4} \int d^4x \; \mathrm{Tr} \left(\mathcal{D}^2 W^\alpha W_\alpha \right) \Big|$$

$$= -\frac{1}{2} \int d^4x \; 2 \; \mathrm{Tr} \left\{ W^\alpha \mathcal{D}^2 W_\alpha - \left(\mathcal{D}^\beta W^\alpha \right) \left(\mathcal{D}_\beta W_\alpha \right) \right\} \Big| \quad .$$

Insertion of the definition (3.1) into the first term immediately gives

$$- \int d^4x \; \mathrm{Tr} \left(W^\alpha \mathcal{D}^2 W_\alpha \right) \Big| = -i \int d^4x \; \mathrm{Tr} \left(\lambda^\alpha \mathcal{D}_{\alpha\dot\alpha} \bar\lambda^{\dot\alpha} \right) \quad .$$

In the second term the field $\mathcal{D}_\beta W_\alpha$ can be decomposed in its symmetric and antisymmetric parts,

$$\mathcal{D}_\beta W_\alpha = \frac{1}{2} \left(\mathcal{D}_\beta W_\alpha + \mathcal{D}_\alpha W_\beta \right) + \frac{1}{2} \, \varepsilon_{\beta\alpha} \, \mathcal{D}^\gamma W_\gamma \quad ; \tag{3.9}$$

the antisymmetric one directly yields a contribution $\frac{1}{2} \int d^4x \; \mathrm{Tr} \; \mathbb{D}^2$ to S_W while use of the Bianchi identities and of their solution leads to the formula

$$\frac{1}{2} \left(\mathcal{D}_\beta W_\alpha + \mathcal{D}_\alpha W_\beta \right) = - \left(\sigma^{ab} \right)_{\alpha\beta} \mathbb{F}_{ab} \tag{3.10}$$

which allows an explicit evaluation of the second term.

The <u>final result</u> is given by

$$S = \int d^4x \left\{ A^\dagger \mathcal{D}^m \mathcal{D}_m A - i \underline{\Psi} \sigma^m \mathcal{D}_m \overline{\underline{\Psi}} - i \frac{\sqrt{2}}{2} \left[A^\dagger (\lambda \cdot \underline{\Psi}) - (\overline{\underline{\Psi}} \cdot \bar{\lambda}) A \right] \right.$$
$$\left. + \frac{1}{2} A^\dagger \mathbb{D} A + F^\dagger F \right\} + \int d^4x \, \mathrm{Tr} \left\{ -\frac{1}{4} \mathbb{V}_{ab} \mathbb{V}^{ab} - i \lambda \sigma^m \mathcal{D}_m \bar{\lambda} + \frac{1}{2} \mathbb{D}^2 \right\} \tag{3.11}$$

and has exactly the same form as the "Lagrangian in the WZ-gauge" of the chiral representation (eq.(2.8)).

(ii) Mass terms

Mass or self interaction terms for the matter field Φ may be included into the action in the familiar way (ref. 154), ch. VII).

In the chiral basis one occasionaly introduces a mass term $m^2 \int d^4\Theta \, V^2$ for the gauge multiplet V : this expression gives a mass to the vector field v_m and introduces the additional degrees of freedom required for a massive multiplet (i.e. a massive scalar and a massive spinor field). An explicit expression for the corresponding term in the real representation is not obvious and probably not of a similarly simple form (for $d=3$ see ref. 57, ch.2.4.c).

(iii) Analogy with the Aharonov-Bohm effect

In the chiral basis the interaction of matter and gauge multiplets is described by the action $\int d^8z \phi^\dagger e^V \phi$ where ϕ obeys the (gauge field independent) chirality condition $\bar{D}_{\dot\alpha}\phi = 0$; in contrast, in the real representation we have a "free" action $\int d^8z \Phi^\dagger \Phi$ for a superfield Φ which is subject to the gauge field dependent condition $\bar{\mathcal{D}}_{\dot\alpha}\Phi = 0$.

An interesting, although quite formal analogy for this situation is provided by the different descriptionsof a well-known effect of non-relativistic quantum mechanics, namely the idealized Aharonov-Bohm (AB) effect*) 166). We recall that the AB effect is concerned with the behaviour of charged particles moving outside (and completely shielded from) an idealized infinitely long solenoid carrying some magnetic flux. To describe this phenomenon one usually considers the interaction Hamiltonian

$$H = \frac{1}{2m} \left(\frac{\hbar}{i} \underline{\nabla} - \frac{e}{c} \underline{A} \right)^2 \tag{3.12}$$

outside the solenoid (supposed to be centered around the z-axis) and wave functions ψ (belonging to the domain of definition of H) which vanish on the boundary of the solenoid and are continuous in its outside : in terms of cylindrical coordinates (r, θ, z) these wave functions and their derivatives w.r.t. θ satisfy

*) Detailed discussions are given in the refs 59), 107).

$$\psi(r, 0, z) = \lim_{\substack{\varepsilon \to 0 \\ \varepsilon > 0}} \psi(r, 2\pi - \varepsilon, z) \tag{3.13}$$

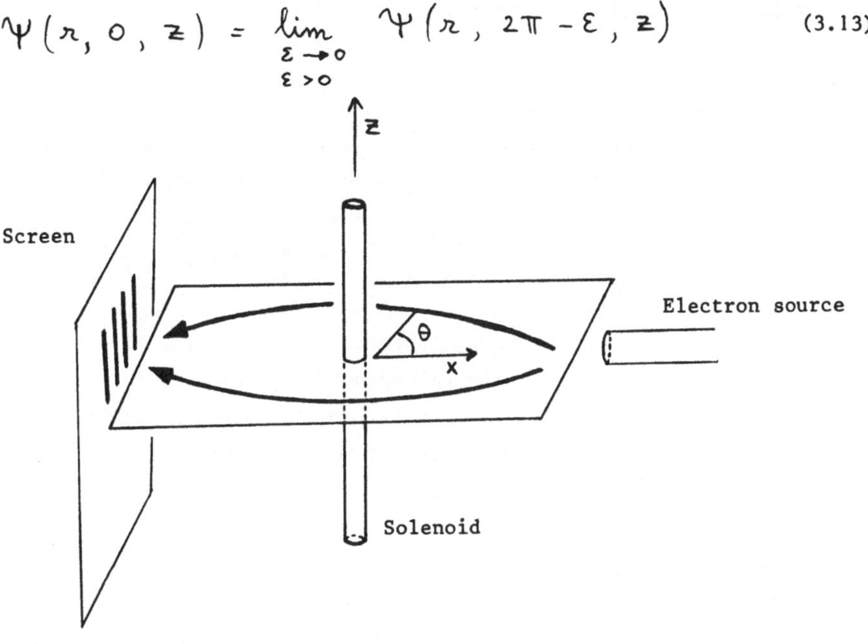

Alternatively (see e.g. ref. 28)) this idealized situation can be described by the "free" Hamiltonian $H = \frac{1}{2m}\left(\frac{\hbar}{i}\underline{\nabla}\right)^2$ and discontinuous wave functions, i.e. more precisely by functions having a discontinuity corresponding exactly to the magnetic flux $\phi \equiv \int \underline{B}.d\underline{f}$ trapped in the AB-solenoid:

$$\psi(r, 0, z) = e^{-\frac{i}{\hbar}\phi} \lim_{\substack{\varepsilon \to 0 \\ \varepsilon > 0}} \psi(r, 2\pi - \varepsilon, z) \tag{3.14}$$

(similarly for the derivatives w.r.t. the variable θ). In our analogy the last equation corresponds to the gauge field dependent condition $\bar{\mathcal{D}}_{\dot{\alpha}}\Phi = 0$.

The equivalence of these two descriptions of the idealized AB-effect is due to the existence of non-equivalent irreducible representations of the quantum-mechanical momentum operator in the non-simply connected space to which the charged particles are confined. Thus it is based on the non-trivial topology of the configuration space. (Of course this does not apply to the equivalence between the real and the chiral representations of SYM-theories.)

We note that a supersymmetric version of the AB-effect has recently been considered in the literature[99)], but the above analogy does not seem to be useful for a description of this phenomenon (or of an eventual non-abelian version[160)]).

IV. BRS DIFFERENTIAL ALGEBRAS IN SYM-THEORIES

In recent years the BRS quantization scheme of YM theories[19)] has been generalized to a large class of related theories, i.e. most notably gauge theories involving diffeomorphisms[168),169),91)] and Weyl transformations (the bosonic string)[170)] or p-form gauge potentials[16)] as well as the supersymmetric versions of these theories, i.e. SYM-theories[57),114)], supergravity[168),171),132),172),145),18)] and superstring theories[77),105),140),173)].

While considering ordinary gauge theories we will first fill the gap between the usual textbook treatment of the BRS symmetry in terms of anticommuting variables and the more abstract differential algebra formulation which is frequently considered in the scientific literature. Thereafter it will be shown how the BRS algebra structure can be incorporated into the geometric framework of SYM-theories. Some aspects of the quantization of these theories will be addressed, in particular the anomaly problem. This part closes with a detailed study of the "BRS component field algebra in the WZ-gauge". Within the framework of supersymmetric theories this algebra has a further reaching significance as a prototype for field dependent differential algebras[*)] and a better geometric insight into the "simple" algebra under consideration should not only be useful for SYM-theories.

IV.1 General introduction to BRS differential algebras

(i) Historical note

The BRS-symmetry has been discovered independently by C. Becchi, A. Rouet and R. Stora[19a)] and by I.V. Tyutin[144)] as an invariance of the effective YM-action. It is the clue to the proof of renormalizability of the theory and the starting point for the algebraic determination of chiral anomalies ; these topics have been systematically studied by BRS and other authors and are reviewed in the refs 20), 133), 111), 15), 136).

(ii) From anticommuting variables to differential algebras

We consider a non-abelian gauge theory with compact matrix group G and Lie algebra $\mathfrak{g}$ (whose generators T^a are supposed to satisfy $[T^a, T^b] = if^{abc}T^c$ and $\mathrm{Tr}(T^aT^b) = \delta^{ab}$). Furthermore let ψ denote a multiplet of fermions coupled to the gauge potential $\underline{\underline{A}}_\mu = A^a_\mu T^a$ and suppose $\underline{\underline{A}}_\mu$ is fixed by the Lorentz gauge condition

---

*) See the references on supergravity and superstring theories given above.

$\partial^\mu A^a_\mu = 0$; then the effective action is given by*)

$$S_{eff} = \int d^4x \left(\mathcal{L}_{class} + \mathcal{L}_{g.f.} + \mathcal{L}_{FPG} \right) \tag{4.1a}$$

where

$$\begin{aligned}
\mathcal{L}_{class} &= -\frac{1}{4} \mathrm{Tr}\left(\underline{\underline{F}}_{\mu\nu} \underline{\underline{F}}^{\mu\nu} \right) + i \bar{\Psi} \gamma^\mu \mathcal{D}_\mu \Psi - m \bar{\Psi} \Psi \\
\mathcal{L}_{g.f.} &= -\frac{1}{2\lambda} \left(\partial^\mu A^a_\mu \right)^2 \qquad (\lambda = \text{gauge parameter}) \\
\mathcal{L}_{FPG} &= i \, \mathrm{Tr} \, \underline{\underline{c}}^\dagger \, \partial^\mu \left(\mathcal{D}_\mu \underline{\underline{c}} \right) \\
\underline{\underline{F}}_{\mu\nu} &= \partial_\mu \underline{\underline{A}}_\nu - \partial_\nu \underline{\underline{A}}_\mu - ig \left[\underline{\underline{A}}_\mu , \underline{\underline{A}}_\nu \right] \qquad (g = \text{coupling constant}) \\
\mathcal{D}_\mu \Psi &= \left(\partial_\mu - ig \, t(\underline{\underline{A}}_\mu) \right) \Psi \\
\mathcal{D}_\mu \underline{\underline{c}} &= \partial_\mu \underline{\underline{c}} - ig \left[\underline{\underline{A}}_\mu , \underline{\underline{c}} \right]
\end{aligned} \tag{4.1b}$$

Here $t(.)$ is a representation of the gauge algebra on the space of matter multiplets ψ and $\underline{\underline{c}} = c^a T^a$ is a complex anticommuting ghost field (with T^a in the adjoint representation of G)[53),41),46)].

It is convenient to replace $\underline{\underline{c}}$ by two real anticommuting fields,

$$\underline{\underline{c}} = \frac{1}{\sqrt{2}} \left(\underline{\underline{\tilde{\omega}}} + i \, \underline{\underline{\omega}} \right)$$

so that the ghost contribution to the effective action becomes

$$\mathcal{L}_{FPG} = i \, \mathrm{Tr} \left[\underline{\underline{\tilde{\omega}}} \, \partial^\mu \left(\mathcal{D}_\mu \underline{\underline{\omega}} \right) \right] . \tag{4.2}$$

(Here we used partial integration, the Lorentz gauge condition and the Grassmann properties of ω^a and $\tilde{\omega}^a$.)

Now the effective action is easily seen to be invariant under the BRS transformations

---

*) For details see for instance ref. 34) or 47),79). Since we do not consider supersymmetric theories in chapter IV.1, we denote the space-time indices by μ, ν as usual.

$$\delta \underset{=}{A}_\mu = \varepsilon \, \mathcal{D}_\mu \underset{=}{\omega}$$
$$\delta \omega^a = -\frac{g\varepsilon}{2} f^{abc} \omega^b \omega^c$$
$$\delta \underset{=}{\tilde{\omega}} = -\frac{i\varepsilon}{\lambda} \partial^\mu \underset{=}{A}_\mu$$
$$\delta \psi = i\varepsilon g \, t(\underset{=}{\omega}) \, \psi \qquad (4.3)$$

where ε is a space-time independent anticommuting parameter*).

To obtain a geometrically more transparent formulation[134),135)] we assign an integer "ghost number" to all the fields: 0 for classical variables $\underset{=}{A}_\mu$ and ψ, +1 for the FP-ghost $\underset{=}{\omega}$ and −1 for the FP-anti-ghost $\underset{=}{\tilde{\omega}}$. Next we note that the transformations (4.3) have the general form

$$\delta \phi = -\varepsilon \cdot (s\phi) \qquad \left(\phi \in \{\underset{=}{A}_\mu, \omega^a, \underset{=}{\tilde{\omega}}, \psi\}\right)$$

with

$$\text{ghost number of } s\phi = \{\text{ghost nb. of } \phi\} + 1 \quad . \qquad (4.4)$$

The fact that the s-operation raises the ghost number by one unit is reminescent of the well-known fact that the exterior differential d raises the form degree of differential forms by one unit ("d acts as an antiderivation of degree one on the exterior algebra").

Since the d-operator is defined in a natural way on the Lie algebra valued forms

$$\omega \equiv i \underset{=}{\omega} \quad , \quad \bar{\omega} \equiv \underset{=}{\tilde{\omega}} \quad , \quad A \equiv A_\mu dx^\mu \equiv -i \underset{=}{A}_\mu dx^\mu \qquad (4.5)$$

and on the multiplet ψ of complex valued functions, we can combine the exterior algebra structure and the BRS symmetry (4.3) (4.4) in the following way : for all fields we add the form degree and the ghost number to obtain a new total degree and we suppose that the algebra generated by A, ω, $\bar{\omega}$ and ψ is graded by this degree. In this spirit all Lie brackets of $\mathfrak{g}$-valued forms correspond to graded commutators,

$$[P, Q] = PQ - (-)^{(\deg P)\cdot(\deg Q)} QP \, ,$$

*) This invariance represents a global symmetry which can be gauged by allowing ε to be space-time dependent and by introducing an anticommuting abelian gauge field $\alpha_\mu(x)$[108),17)].

for instance

$$[A, \omega] = A\omega + \omega A$$

$$[\omega, \omega] = \omega\omega + \omega\omega \quad .$$

On the so-defined algebra the operators d and s act as antiderivations of degree 1, which in addition are required to satisfy $sd + ds = 0$:

$$
\begin{aligned}
sA &= -\mathcal{D}\omega = -\left(d\omega + [A, \omega]\right) \\
s\omega &= -\frac{1}{2}[\omega, \omega] \\
s\bar{\omega} &= \frac{1}{\lambda} d^*A = -\frac{1}{\lambda}\left(\partial^\mu A_\mu\right) \\
s\psi &= -t(\omega)\,\psi
\end{aligned}
\tag{4.6}
$$

Here we assumed that $g = 1$ and we denoted the codifferential of d by d^* [21]. The global sign in the first equation was chosen so as to have

$$s^2 A = 0$$

as a consequence of the relations (4.6) and of the properties of d and s ; we also have

$$s^2\omega = s^2\psi = 0 \quad .$$

These relations reflect the closure and the Jacobi identity of the gauge algebra [19].

We note that the s-variation of the field strength (curvature 2-form) $F = dA + \frac{1}{2}[A,A] = dA + AA$ can be directly derived from the one of A by using the basic properties of s and d :

$$sF = [F, \omega] = F\omega - \omega F \quad . \tag{4.7}$$

Thus, on the classical fields A, F and ψ the s-operation corresponds to an infinitesimal gauge transformation with parameter ω. On the other hand the relation for ω has exactly the same form as the Maurer-Cartan equation of a Lie group (with d replaced by s) ; this similarity can be given a precise mathematical meaning by an appropriate algebraic characterization of the FP-ghost ω [136)*)].

---

*) Different geometric interpretations of ω and of its s-variation have been given and discussed by several authors (refs 133),93),141),96),24),25),135),164),15),147)).

In the next section we will show that the introduction of an auxiliary field allows to reformulate the effective action and the BRS transformations in such a way that the s-operator is nilpotent on all the fields on which it is defined. This description also represents a canonical method for incorporating the FP-antighost field $\bar{\omega}$. Therefore one often restricts oneself to the first two equations in (4.6) : their mathematical structure will be discussed in section (V). Here we only observe that they are equivalent to the component field algebra defined by the relations

$$\begin{aligned} s A_\mu &= \mathcal{D}_\mu \omega = \partial_\mu \omega + A_\mu \omega - \omega A_\mu \\ s\omega &= -\frac{1}{2}[\omega,\omega] = -\omega\omega \\ s^2 &= 0 = [s, \partial_\mu] \quad . \end{aligned} \tag{4.8}$$

On this algebra the operator s acts an an antiderivation of degree 1 whereas the partial differential operators ∂_μ act as derivations which commute with s. The (nilpotent) s-variation of A_μ follows from the one of A by use of the following differentiation rule of a function with ghost number one:

$$d\omega = -(\partial_\mu \omega)\, dx^\mu . \tag{4.9}$$

(Note that for a 1-form $\alpha = \alpha_\nu\, dx^\nu$ one has formally (see also eq.(4.70)):

$$d\alpha = \partial_\mu \alpha_\nu\, dx^\mu dx^\nu = -\partial_\mu[\alpha_\nu dx^\nu]\, dx^\mu = -(\partial_\mu \alpha)\, dx^\mu .)$$

(iii) The "canonical formulation"[161),89),136)]

First we note that with (4.5),(4.8) the ghost Lagrangian (4.2) can be rewritten as

$$\begin{aligned} \mathcal{L}_{FPG} &= i\, \mathrm{Tr}\, \underline{\underline{\tilde{\omega}}}\, \partial^\mu(\mathcal{D}_\mu \underline{\underline{\omega}}) = \mathrm{Tr}\, \bar{\omega}\, \partial^\mu(\mathcal{D}_\mu \omega) = \mathrm{Tr}\, \bar{\omega}\, \partial^\mu(sA_\mu) \\ &= \mathrm{Tr}\, \bar{\omega}\, s(\partial^\mu A_\mu) = \mathrm{Tr}\, \bar{\omega}\, (s\mathcal{G}) \end{aligned} \tag{4.10}$$

where

$$\mathcal{G}^a(A) = \partial^\mu A^a_\mu = 0 \tag{4.11}$$

represents the gauge fixing condition. The latter can be incorporated into the effective action by the Lagrange multiplier method ; since $\mathcal{G}$ is a $\mathfrak{g}$-valued function we have to consider a $\mathfrak{g}$-valued Lagrange multiplier field $b(x) = b^a(x)T^a$ (with T^a in the adjoint representation of G):

$$\mathcal{L}_{g.f.} := \mathrm{Tr}\, b\mathcal{G} = b^a \mathcal{G}_a \quad .$$

The commuting scalar field b has been introduced more or less independently and explicitly by several authors[137),100),92),161),20),89)] and is usually referred to as the auxiliary field ; it deserves this name since it has canonical dimension 2 and thus cannot acquire kinetic energy in any renormalizable Lagrangian[15)].

Now we want to define the BRS transformations of $\bar{\omega}$ and b in such a way that they leave the effective action invariant and such that $s^2\bar{\omega} = 0 = s^2 b$. Suppose s has already been defined as a nilpotent operator. Then the Lagrangian

$$\mathcal{L}_{g.f.} + \mathcal{L}_{FPG} = \mathrm{Tr}\left[b\mathcal{G} + \bar{\omega}(s\mathcal{G}) \right]$$

is certainly s-invariant, if it has the form s(.) ; thus we are led to the definition of the (nilpotent) s-operations

$$s\bar{\omega} = -b \quad , \quad sb = 0$$

which imply

$$\begin{aligned}\mathcal{L}_{g.f.} + \mathcal{L}_{FPG} &= \mathrm{Tr}\left[-(s\bar{\omega})\mathcal{G} + \bar{\omega}(s\mathcal{G}) \right] \\ &= -\mathrm{Tr}\; s(\bar{\omega}\mathcal{G}) \qquad \text{(Recall : deg } \bar{\omega} = -1).\end{aligned}$$

To summarize : The effective action

$$S_{eff} = \int d^4x \left\{ \mathcal{L}_{class}(A,\psi) + \mathrm{Tr}\left[b\cdot\mathcal{G}(A) + \bar{\omega}\, s\mathcal{G} + b^a M_{ac} b^c \right] \right\} \tag{4.12}$$

(where M_{ac} represents a constant numerical matrix) is invariant under the BRS transformations

$$\begin{aligned} sA &= -(d\omega + [A,\omega]) \\ s\omega &= -\tfrac{1}{2}[\omega,\omega] \\ s\bar{\omega} &= -b \\ sb &= 0 \\ s\psi &= -t(\omega)\psi \\ s^2 &= 0 = sd + ds \end{aligned} \tag{4.13}$$

This algebra which is characterized by a nilpotent s-operation is called the BRS differential algebra. If we disregard the matter fields it can be decomposed into two pieces : the geometric part given by the first two equations and a contractible part[94),167)] defined by the following two equations.

The formulation of the effective action and of the BRS algebra given in the last section can be directly recovered from the present one for $M_{ac} = \frac{\lambda}{2} \delta_{ac}$. In this case the equations of motion for b^a have the form $\lambda . b_a = - \mathcal{G}_a$ and insertion of this expression in the action (4.12) yields

$$
\begin{aligned}
S_{eff} &= \int d^4x \left\{ \mathcal{L}_{class} - \frac{1}{\lambda} \mathcal{G}^a \mathcal{G}_a + \bar{\omega}^a (s\mathcal{G})_a + \frac{1}{2\lambda} \mathcal{G}^a \mathcal{G}_a \right\} \\
&= \int d^4x \left\{ \mathcal{L}_{class} - \frac{1}{2\lambda} \mathcal{G}^a \mathcal{G}_a + \bar{\omega}^a (s\mathcal{G})_a \right\} \\
&= \int d^4x \left\{ \mathcal{L}_{class} - \frac{1}{2\lambda} (\partial^\mu A^a_\mu)^2 + \bar{\omega}^a \partial_\mu (s A^\mu_a) \right\}
\end{aligned}
\tag{4.14}
$$

which coincides with our original expression for S_{eff}.

Similarly insertion of $b = -\frac{1}{\lambda} \mathcal{G}$ in $s\bar{\omega} = -b$ reproduces the original s-variation of $\bar{\omega}$; the latter is not nilpotent since the equations of motion for b are not BRS-invariant ($sb = 0$, but $s\mathcal{G}(A) \neq 0$).

In the Feynman gauge one chooses $\lambda = 1$ whereas the Landau gauge ($\lambda = 0$) is implemented by dropping the last term in the Lagrangian (4.12). This choice is generally made when considering a background gauge fixing procedure[170].

(iv) About anti-BRS transformations

In analogy to the BRS operator s which increases the ghost number by one unit, one can introduce an "anti-BRS operator" $\bar{s}$ which decreases this number by the same amount, but which otherwise has the same mathematical properties as s (antiderivation of degree 1, $\bar{s}d + d\bar{s} = 0$, $\bar{s}(A_\mu dx^\mu) = -(\bar{s} A_\mu)dx^\mu$). The combined BRS/anti-BRS transformations are as follows[37),106),14)]:

$$
\begin{aligned}
sA &= -D\omega & \bar{s}A &= -D\bar{\omega} \\
s\omega &= -\tfrac{1}{2}[\omega, \omega] & \bar{s}\bar{\omega} &= -\tfrac{1}{2}[\bar{\omega}, \bar{\omega}] \\
s\bar{\omega} &= b & \bar{s}\omega &= -[\bar{\omega}, \omega] - b \\
sb &= 0 & \bar{s}b &= -[\bar{\omega}, b]
\end{aligned}
$$

$$
0 = s^2 = \bar{s}^2 = s\bar{s} + \bar{s}s = sd + ds = \bar{s}d + d\bar{s} . \tag{4.15}
$$

An admissible s and $\bar{s}$ invariant quantum Lagrangian is given by[37),14)]

$$\mathcal{L} = \mathrm{Tr}\left[-\frac{1}{4}(F_{\mu\nu})^2 - \frac{1}{2} s\bar{s}\left(A_\mu^2 - \lambda\bar{\omega}\omega\right)\right]$$

$$= \mathrm{Tr}\left[-\frac{1}{4}(F_{\mu\nu})^2 + b\left(\partial_\mu A^\mu + \frac{\lambda}{2}[\bar{\omega},\omega]\right) - \bar{\omega}\,\partial^\mu(\mathcal{D}_\mu\omega) + \frac{\lambda}{4}[\bar{\omega},\omega]^2 + \frac{\lambda}{2}b^2\right].$$

By the redefinition $b' = b + \frac{1}{2}[\omega,\omega]$ and the subsequent elimination of b' (by means of its equations of motion) this Lagrangian can be cast into the form

$$\mathcal{L} = \mathrm{Tr}\left[-\frac{1}{4}(F_{\mu\nu})^2 - \frac{1}{2\lambda}(\partial^\mu A_\mu)^2 - \frac{1}{2}\bar{\omega}\,\partial^\mu(\mathcal{D}_\mu\omega) + \frac{1}{2}\omega\,\partial^\mu(\mathcal{D}_\mu\bar{\omega}) + \frac{\lambda}{8}[\bar{\omega},\omega]^2\right] \tag{4.16}$$

Comparing the BRS-variations of A and ω with the anti-BRS-transformations of A and $\bar{\omega}$ as well as the quantum Lagrangians (4.14),(4.16) one remarks that the anti-BRS symmetry essentially represents a copy of the BRS symmetry with the roles of ω and $\bar{\omega}$ exchanged. (Moreover eq.(4.16) tells us that the $\bar{s}$-invariance fixes a certain class of gauges, namely the linear ones.) Eventually there is no known model for which the $\bar{s}$-invariance provides any information which does not yet follow from its s-invariance. For this reason one usually disregards the anti-BRS symmetry*).

(v) <u>BRS differential algebras</u>

In order to define properly the anomalous terms of the quantum theory, one has to characterize the BRS algebra in more mathematical terms as a particular bigraded differential algebra.

First we note that the definition of an ordinary differential algebra represents an abstraction of the characteristic properties of the algebra of differential forms on $\mathbb{R}^n$, i.e. the exterior algebra $\Omega^*(\mathbb{R}^n) = \bigoplus_{p\geq 0} \Omega^p(\mathbb{R}^n)$. By definition[63),64)] a pair $(\mathfrak{A}^*, d)$ is a <u>graded differential algebra</u> (or differential algebra for short), if

(i) $\mathfrak{A}^*$ is an associative algebra over the field $\mathbb{Q}$, $\mathbb{R}$ or $\mathbb{C}$ and $\mathfrak{A}^*$ is positively graded (i.e. $\mathfrak{A}^*$ as a vector space admits a direct sum decomposition

$$\mathfrak{A}^* = \bigoplus_{p\geq 0} \mathfrak{A}^p$$

*) Yet it is interesting that one can rederive the well-known properties of YM-theories by postulating both BRS and anti-BRS invariance of the quantum Lagrangian and of the functional integral measure, see ref. 14), and the review of L. Baulieu[15)].

and the algebra product satisfies $\mathfrak{a}^p . \mathfrak{a}^q \subset \mathfrak{a}^{p+q}$)

(ii) d is an antiderivation of degree 1 (i.e. $d = \mathfrak{a}^* \to \mathfrak{a}^*$ is a linear map with $d(\mathfrak{a}^p) \subset \mathfrak{a}^{p+1}$ for all $p \geq 0$ and d satisfies the anti-Leibniz rule

$$d(\alpha\cdot\beta) = (d\alpha)\cdot\beta + (-1)^p \alpha\cdot(d\beta) \quad \text{for} \quad \alpha \in \mathfrak{a}^p, \beta \in \mathfrak{a}^*)$$

(iii) d is nilpotent (i.e. $d^2 = 0$)

Since many differential algebras of interest have this property, one often requires in addition that

(iv) $\mathfrak{a}^*$ is graded commutative (i.e.

$$\alpha\cdot\beta = (-1)^{p\cdot q}\, \beta\cdot\alpha \qquad \text{for} \quad \alpha \in \mathfrak{a}^p, \beta \in \mathfrak{a}^q)$$

The BRS algebra is characterized by two gradings and two differentials and therefore we have to generalize the previous definition[44] : a pair $(\mathfrak{a}^*, d)$ is a <u>bigraded differential algebra</u>, if

(i) $\mathfrak{a}^*$ is a bigraded algebra (i.e. $\mathfrak{a}^*$ is an associative algebra which admits a direct sum decomposition

$$\mathfrak{a}^* = \bigoplus_{p,q \geq 0} \mathfrak{a}^{(p,q)}$$

and the algebra product satisfies $\mathfrak{a}^{(p,q)} . \mathfrak{a}^{(p',q')} \subset \mathfrak{a}^{(p+p',q+q')}$).

(ii) $d = d^{(1,0)} + d^{(0,1)}$ where $d^{(1,0)}$ and $d^{(0,1)}$ are antiderivations which raise, respectively, the first and the second degree by 1 (i.e. $d^{(1,0)}$, $d^{(0,1)}$ are linear maps from $\mathfrak{a}^*$ to $\mathfrak{a}^*$ and satisfy

$$d^{(1,0)}\, \mathfrak{a}^{(p,q)} \subset \mathfrak{a}^{(p+1,q)} \quad , \quad d^{(0,1)}\, \mathfrak{a}^{(p,q)} \subset \mathfrak{a}^{(p,q+1)}$$

and

$$\left.\begin{aligned} d^{(1,0)}(\alpha\cdot\beta) &= (d^{(1,0)}\alpha)\cdot\beta + (-1)^{p+q}\alpha\cdot(d^{(1,0)}\beta) \\ d^{(0,1)}(\alpha\cdot\beta) &= (d^{(0,1)}\alpha)\cdot\beta + (-1)^{p+q}\alpha\cdot(d^{(0,1)}\beta) \end{aligned}\right\} \text{ for } \alpha \in \mathfrak{a}^{(p,q)}, \beta \in \mathfrak{a}^*)$$

(iii) $0 = (d^{(1,0)})^2 = (d^{(0,1)})^2 = d^{(1,0)}d^{(0,1)} + d^{(0,1)}d^{(1,0)}$.

$\mathfrak{a}^*$ is graded commutative, if

$$\alpha \cdot \beta = (-1)^{(p+q)(p'+q')} \beta \cdot \alpha \quad \text{for} \quad \alpha \in \mathcal{O}\!\iota^{(p,q)}, \quad \beta \in \mathcal{O}\!\iota^{(p',q')}$$

Again a simple example for such an algebra is provided by the de Rham algebra $\Omega^*(\mathbb{R}^{n+m})$. On this simply graded algebra we can introduce a bidegree (p = degree in $dx^1,\ldots,dx^n$; q = degree in $dx^{n+1},\ldots, dx^{n+m}$):

$$\Omega^*(\mathbb{R}^{n+m}) = \bigoplus_{p,q\geq 0} \Omega^{(p,q)}(\mathbb{R}^{n+m}) . \tag{4.17}$$

Thus an element $\omega \in \Omega^{(p,q)}(\mathbb{R}^{n+m})$ has the standard form

$$\omega(x) = \sum_{\substack{1\leq i_1<\ldots<i_p\leq n \\ n+1\leq j_1<\ldots<j_q\leq n+m}} \omega_{i_1\ldots j_q}(x)\, dx^{i_1}\wedge\ldots\wedge dx^{i_p}\wedge dx^{j_1}\wedge\ldots\wedge dx^{j_q}$$

If the total differential $d = \sum_{i=1}^{n+m} dx^i\partial_i$ is similarly decomposed,

$$d = d^{(1,0)} + d^{(0,1)} = \sum_{i=1}^{n} dx^i\partial_i + \sum_{j=n+1}^{n+m} dx^j\partial_j$$

the bigraded algebra (4.17) equipped with this differential defines a bigraded differential algebra.

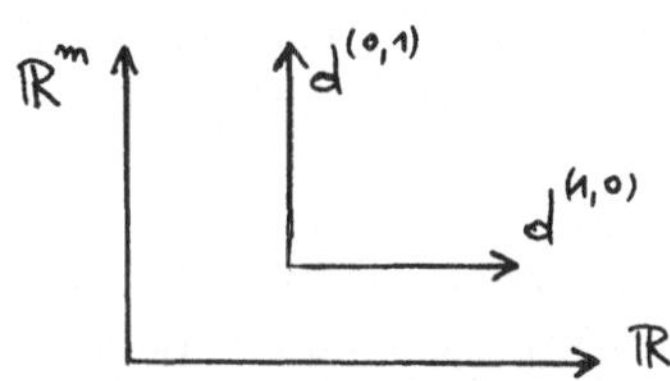

A less trivial example is given by the "universal BRS algebra $A(\mathfrak{g})$ associated to a finite dimensional Lie algebra $\mathfrak{g}$"[43]. Consider the bigraded commutative algebra generated by some abstract $\mathfrak{g}$-valued objects $A = A^aT^a$, $F = F^aT^a$, $\omega = \omega^aT^a$, $\varphi = \varphi^aT^a$ of respective bidegrees (1,0), (2,0), (0,1), (1,1)*). On these generators we define the total differential $d^{(1,0)} + d^{(0,1)} = d + s$ by**)

$$\begin{aligned} dA &= F - \tfrac{1}{2}[A,A] & sA &= -\varphi - [A,\omega] \\ dF &= [F,A] & sF &= [F,\omega] \\ d\omega &= \varphi & s\omega &= -\tfrac{1}{2}[\omega,\omega] \\ d\varphi &= 0 & s\varphi &= [\varphi,\omega] \end{aligned} \tag{4.18}$$

---

*) An arbitrary element of $A(\mathfrak{g})$ is a linear combination of products of generators.

**) Note that $dF = [F,A]$ formally coincides with the Bianchi identity of the curvature 2-form.

Here [,] denotes the natural Lie bracket on $A(\mathfrak{g})$, e.g.

$$[A^a T^a, \omega^b T^b] = A^a \cdot \omega^b \, [T^a, T^b] \quad .$$

The action of the antiderivations d and s on the generators of $A(\mathfrak{g})$ extends uniquely to the whole algebra ; as one easily verifies, d and s satisfy the relations $d^2 = s^2 = ds + sd = 0$.

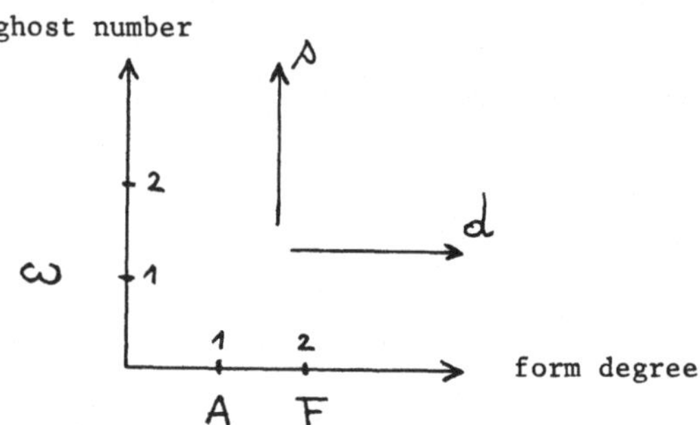

Our original BRS algebra (4.13),(4.7) is a realization of the universal BRS algebra for which $\mathfrak{g}$ is the Lie algebra of the symmetry group and for which A^a, F^a, ω^a, φ^a are real valued differential forms on $\mathbb{R}^4$ while d corresponds to the usual exterior differential.

For further mathematical developments and their application to the determination of all possible anomalous terms in YM-theories we refer to the original articles[19b),94),43),84)] and to the comprehensive reviews of R. Stora[136)] and M. Dubois-Violette[44)].

(vi) Including space-time transformations

When discussing the "BRS component field algebra in the WZ-gauge" we do not only have to deal with gauge transformations, but also with space-time translations and susy transformations. In the superspace approach these are described by a supervector field $\xi = \xi^M \partial_M$ on which the BRS operator acts in the following way

$$s\,\xi = -\frac{1}{2}[\xi, \xi] \qquad \left(\Leftrightarrow \quad s\,\xi^M = -\xi^N (\partial_N \xi^M)\right) \tag{4.19}$$

On the other hand the exterior differential does not act on ξ and therefore one has to modify slightly the definition of the BRS differential algebra structure in order to incorporate properly the space-time transformations. This can be done in the following way[*),91),136),120)].

[*)] The argumentation applies to ordinary manifolds as well as superspaces.

If ξ is a vector field on the space M, the interior product i_ξ is an antiderivation of degree -1 on the exterior algebra of M and $L_\xi = i_\xi d + d i_\xi = \{i_\xi, d\}$ is a derivation on the same algebra. When turning ξ into a FP-ghost with ghost number one, the operator i_ξ becomes a derivation which satisfies

$$\left(i_\xi\right)^{k+1} \alpha_k = 0 \tag{4.20}$$

if α_k has a form degree k and the Lie derivative

$$L_\xi \equiv i_\xi d + (-)^{\deg \xi} d i_\xi$$

$$= i_\xi d - d i_\xi$$

becomes an antiderivation of degree 1. Note that $L_{s\xi} = i_{s\xi} d + d i_{s\xi}$ again represents a derivation since $s\xi$ has an even ghost number.

We now introduce the ghost vector field ξ in the bigraded BRS differential algebra by assuming that i_ξ acts on this algebra as a derivation with the properties

$$[i_\xi, i_\eta] = 0 \ , \quad [s, i_\xi] = i_{s\xi} \ , \quad [i_\xi, d] = L_\xi$$

$$\{s, L_\xi\} = L_{s\xi} \ , \quad [L_\xi, i_\eta] = i_{[\xi,\eta]} \ , \quad [L_\xi, d] = 0$$

$$[L_\xi, L_\eta] = L_{[\xi,\eta]} \ . \tag{4.21}$$

The transformation law (4.19) can be incorporated in this general setting by the relation

$$i_{s\xi} = -\frac{1}{2} i_{[\xi,\xi]} \ . \tag{4.22}$$

We conclude with a comment concerning the possible anomalous terms associated to a gauge theory involving space-time diffeomorphisms. While all such terms have been determined for ordinary gauge theories by BRS[19b)] and by M. Dubois-Violette, M. Talon and C. Viallet[43)], this work has not been generalized so as to incorporate vector fields ; though[*)] many particular representatives have been obtained by the descent equations method which will be described below.

---

*) See references 91), 136) and references therein.

(vii) Associated Lie algebras and characterization of anomalies

In the subsequent chapters we will frequently consider the correspondence between Lie algebras and BRS differential algebras ; also we will refer to the cohomology of Lie algebras and the cohomological definition of anomalies. This terminology might frighten the unprepared reader, but we hope that the following elementary account will convince him that the involved concepts are fairly simple and really useful. Our presentation is based on the work of R. Stora and B. Zumino[134),135),136),164),183)] and is restricted to the case of a gauge theory with trivial principal fibre bundle (over 4-dimensional Minkowski space) ; the general case is treated in the references 94),84).

As structure group we consider a compact Lie group G. The corresponding gauge group $\mathcal{G}$ consists of all smooth functions from space-time M to G :

$$\mathcal{G} = \{ g : M \longrightarrow G \}$$

The Lie algebra Lie $\mathcal{G}$ of $\mathcal{G}$ can be identified in a natural way with the vector space of maps from space-time to Lie G with a Lie bracket defined by

$$[\omega_1 , \omega_2](x) = [\omega_1(x) , \omega_2(x)] \qquad (\omega_1, \omega_2 \in \mathrm{Lie}\,\mathcal{G})$$

The YM connection A is a Lie G-valued one form on M which transforms under $g \in \mathcal{G}$ as

$$A(x) \mapsto {}^{g}\!A(x) = g(x)^{-1} A(x)\, g(x) + g^{-1}(x)\, dg(x)$$

respectively, under $\omega \in$ Lie $\mathcal{G}$ as

$$\delta_\omega A = D\omega = (d\omega + [A, \omega]) \quad . \tag{4.23a}$$

For a matter field ψ taking its values in the vector space of a linear representation T of G we have the transformation law

$$\psi(x) \mapsto {}^{g}\psi(x) = T\left(g(x)^{-1}\right)\psi(x) \qquad \text{for} \qquad g \in \mathcal{G}$$

respectively,

$$\delta_\omega \psi = - t(\omega)\, \psi \qquad \text{for} \qquad \omega \in \mathrm{Lie}\,\mathcal{G} \quad . \tag{4.23b}$$

Here t is the representation of Lie $\mathcal{G}$ induced by T.

Let $\Gamma(A,\psi)$ denote the vector space of all functionals of A and ψ and let $\Gamma_{loc}(A,\psi)$ be the subspace of local functionals of A and ψ, i.e. functionals of the form

$$\int_M dx \; P(A_\mu, \psi, \partial_\nu A_\mu, \partial_\nu \psi)(x)$$

where P is a locally gauge invariant polynomial depending on A_μ, ψ and their derivatives to finite order*).

The infinitesimal transformations (4.23) induce a variation of the functionals of A and ψ which can be described by the functional derivatives (Ward operators)

$$W(\omega) = \int_M dx \left[(\delta_\omega A) \frac{\delta}{\delta A} + (\delta_\omega \psi) \frac{\delta}{\delta \psi} \right]$$

$$\equiv \int_M dx \left[(\delta_\omega A_\mu^a)(x) \frac{\delta}{\delta A_\mu^a(x)} + (\delta_\omega \psi^i)(x) \frac{\delta}{\delta \psi^i(x)} \right] \quad . \tag{4.24}$$

As a consequence of their definition these linear operators define a representation of the Lie algebra Lie $\mathcal{G}$ on the vector space $\Gamma(A,\psi)$:

$$\left[W(\omega) , W(\omega') \right] = W\left([\omega, \omega'] \right) \tag{4.25}$$

Now the (BRS) differential algebra associated to the Lie algebra $\{W(\omega)\}$ given by (4.23-4.25) is defined by the relations

$$\begin{aligned} s A &= -\mathcal{D}\omega = -\left(d\omega + [A, \omega] \right) \\ s\omega &= -\tfrac{1}{2} [\omega, \omega] \\ s\psi &= -t(\omega)\psi \end{aligned} \tag{4.26}$$

---

*) An example for a non-local functional is given by

$$\int dx \; \bar{\psi}(x) \sum_{n=0}^{\infty} \frac{1}{n!} (a^\mu \partial_\mu)^n \psi(x) = \int dx \; \bar{\psi}(x) \psi(x+a)$$

or by

$$\int dx \int dy \; \bar{\psi}(y) \, G(x-y) \, \psi(x)$$

where $G(x-y)$ represents an integral kernel that is not of the form $\delta^{(4)}(x-y)$ times a polynomial in ∂_μ (e.g. $G(x-y) = \frac{1}{\square}(x-y)$ defined by

$$\square_x G(x-y) = \delta^{(4)}(x-y) \qquad).$$

Such non-local interaction terms cannot occur in the Lagrangian of a local field theory.

Here ω represents a FP-ghost with total degree one.

Anomalies correspond to a certain breaking of the classical symmetries at the quantum level. They can be defined in terms of the basic notions we just introduced.

The gauge invariance of the classical action $S(A,\psi)$ is expressed by the identity

$$W(\omega)\ S(A,\psi) = 0$$

The "quantum action" is given by the generating functional of one-particle-irreducible Green functions (the vertex functional) $S_{quant}(A,\psi)$ which we assume to be renormalized in the one-loop approximation. In general, in the presence of chiral fermions the quantum action is not invariant under $\mathcal{G}$:

$$W(\omega)\ S_{quant}(A,\psi) = \Delta(\omega, A) \quad \text{(Anomalous Ward identity)} \tag{4.27}$$

(Here $W(\omega)$ is the Ward operator (4.24) expressed in terms of the renormalized fields A,ψ). Since $W(\omega)$ is linear in ω the so-called Adler-Bardeen anomaly $\Delta(\omega,A)$ is also linear in ω. Furthermore it is a local functional in ω and A, i.e. $\Delta(\omega,A) = \int_M \alpha(\omega,A)$. The 4-form α is only defined up to total derivatives because $\int_M d\chi = 0$.

As a consequence of the commutation relations (4.25) the anomaly $\Delta(\omega,A)$ necessarily satisfies the Wess-Zumino consistency conditions[151]:

$$\begin{aligned} 0 &= \left\{ \left[W(\omega), W(\omega')\right] - W\left([\omega,\omega']\right) \right\} S_{quant}(A,\psi) \\ &= W(\omega)\Delta(\omega',A) - W(\omega')\Delta(\omega,A) - \Delta\left([\omega,\omega'],A\right) . \end{aligned} \tag{4.28}$$

A trivial solution of this functional differential equation can be directly given:

$$\Delta(\omega,A) = W(\omega)\ \tilde{S}(A) \tag{4.29}$$

with $\tilde{S}(A)$ being a local functional of A. Such a solution is not very interesting, since it just amounts to a redefinition of the quantum action,

$$S_{quant}(A,\psi) \longrightarrow S_{quant}(A,\psi) + \tilde{S}(A)$$

and corresponds to a change of the renormalization prescription which does not introduce a dependence on ψ. Thus one defines the anomaly as a solution of the WZ-consistency equations (4.28) which cannot be written as a variation of a local functional of the basic fields (A,ψ). This characterizes the anomaly as an element of the cohomology group $H^1_{loc}(\text{Lie }\mathcal{G}, \Gamma_{loc}(A))$ which we will now introduce.

The cohomology groups of Lie algebras are defined in complete analogy to the de Rham cohomology groups. Therefore we first recall the corresponding notions[150]. A real valued p-form on a manifold M is a p-linear and totally antisymmetric map on the vector space $\mathfrak{X}(M)$ of vector fields on M :

$$\alpha : \mathfrak{X}(M) \times \ldots \times \mathfrak{X}(M) \longrightarrow C^\infty(M)$$

$$(X_1, \ldots, X_p) \longmapsto \alpha(X_1, \ldots, X_p).$$

The differential d can be defined on the exterior algebra $\Omega^*(M) = \bigoplus_{p \geq 0} \Omega^p(M)$ by

$$(d\alpha)(X_1, \ldots, X_{p+1}) = \sum_{i=1}^{p+1} (-1)^{i+1} X_i \left[\alpha(X_1, \ldots, \hat{X}_i, \ldots, X_{p+1})\right]$$
$$+ \sum_{1 \leq i < j \leq p+1} (-1)^{i+j} \alpha\left([X_i, X_j], X_1, \ldots, \hat{X}_i, \ldots, \hat{X}_j, \ldots, X_{p+1}\right)$$

for $\alpha \in \Omega^p(M)$ and $X_1, \ldots, X_{p+1} \in \mathfrak{X}(M)$.

Let $Z^p(M)$ and $B^p(M)$ denote the set of closed and exact p-forms on M :

$$Z^p(M) = \{\alpha \in \Omega^p(M) \;/\; d\alpha = 0\}$$

$$B^p(M) = \{\alpha \in \Omega^p(M) \;/\; \alpha = d\beta \text{ for some } \beta \in \Omega^{p-1}(M)\}$$

Since $d^2 = 0$ one has $B^p(M) \subset Z^p(M)$ and therefore one can introduce the p-th de Rham cohomology group

$$H^p(M) = Z^p(M) / B^p(M)$$

When using the general algebraic terminology[150], one calls d a coboundary operator while the elements of $\Omega^p(M)$, $Z^p(M)$, $B^p(M)$ and $H^p(M)$ are referred to as cochains, cocycles, coboundaries and cohomology classes, respectively.

Next we define the corresponding objects of the Lie algebra cohomology[192),70)]. We consider a Lie group G with Lie algebra Lie G and a linear representation T of G on a finite dimensional vector space V (as well as the induced representation t of Lie G).

For $p \in \mathbb{N}$ let C^p (Lie G, V) denote the vector space of p-linear alternate forms on Lie G, with values in V :

$$f_p : \mathcal{L}ie\, G \times \ldots \times \mathcal{L}ie\, G \longrightarrow V$$
$$(X_1, \ldots, X_p) \longmapsto f_p(X_1, \ldots, X_p) \,.$$

For $p = 0$ we set $C^o(\text{Lie } G, V) := V$. The elements f_p are called p-cochains with values in V.

On $C^*(\text{Lie } G, V) \equiv \bigoplus_{p \geq o} C^p(\text{Lie } G, V)$ we define the coboundary operator δ by

$$\delta : C^p(\mathcal{L}ie\, G, V) \longrightarrow C^{p+1}(\mathcal{L}ie\, G, V)$$
$$f_p \longmapsto \delta f_p$$

$$(\delta f_p)(X_1, \ldots, X_{p+1}) = \sum_{i=1}^{p+1} (-1)^{i+1}\, t(X_i) \cdot f_p(X_1, \ldots, \hat{X}_i, \ldots, X_{p+1})$$
$$+ \sum_{1 \leq i < j \leq p+1} (-1)^{i+j} f_p([X_i, X_j], X_1, \ldots, \hat{X}_i, \ldots, \hat{X}_j, \ldots, X_{p+1}) \tag{4.30a}$$

where $X_1, \ldots, X_{p+1} \in \text{Lie } G$. For instance

$$(\delta f_o)(X) = t(X)\, f_o \tag{4.30b}$$

and

$$(\delta f_1)(X, Y) = t(X)\, f_1(Y) - t(Y)\, f_1(X) - f_1([X, Y]) \,. \tag{4.30c}$$

From the definition of δ it follows that $\delta^2 = 0$. As for the de Rham complex we define the vector space of cocycles, coboundaries and cohomology classes by

$$Z^p(\mathcal{L}ie\, G, V) = \{ f_p \,/\, \delta f_p = 0 \}$$
$$B^p(\mathcal{L}ie\, G, V) = \{ f_p \,/\, f_p = \delta f_{p-1} \text{ for some } f_{p-1} \in C^{p-1} \}$$
$$H^p(\mathcal{L}ie\, G, V) = Z^p(\mathcal{L}ie\, G, V) \big/ B^p(\mathcal{L}ie\, G, V) \,. \tag{4.30d}$$

We now come back to the cohomological characterization of anomalies and consider the substitutions

$$\begin{array}{ccc} \text{Lie } G & \longrightarrow & \text{Lie } \mathcal{G} \\ V & \longrightarrow & \Gamma_{loc}(A) \\ t(\chi) & \longrightarrow & W(\omega) \\ \text{degree } p & \longrightarrow & \text{degree in } \omega \text{ (ghost number)} \end{array}$$

Comparing the relations (4.28),(4.29) and (4.30b-c) we conclude that an <u>anomaly</u> $\Delta(\omega,A)$, interpreted as a linear map

$$\Delta(\cdot\,, A) : \mathcal{L}ie\, \mathcal{G} \longrightarrow \Gamma_{loc}(A)$$
$$\omega \longmapsto \Delta(\omega, A)$$

<u>is a representative of a non-trivial element of</u> $H^1(\text{Lie}\,\mathcal{G}, \Gamma_{loc}(A))$:

$$0 = \Big(\delta\Delta(\cdot\,, A)\Big)(\omega, \omega') = W(\omega)\,\Delta(\omega', A) - W(\omega')\,\Delta(\omega, A) - \Delta\big([\omega,\omega'], A\big) \tag{4.31}$$

Properlý speaking the anomaly $\Delta(.,A)$ belongs to the <u>local</u> cohomology group $H^1_{loc}(\text{Lie}\,\mathcal{G}, \Gamma_{loc}(A))$, since it is always of the form

$$\omega \longmapsto \mathrm{Tr} \int dx\, \omega(x)\; \Delta\big(A_\mu(x), \partial_\nu A_\mu(x)\big)$$

or it can be directly cast into this form by partial integration[19b].

As pointed out by J. Dixon and R. Stora[134] equation (4.31) can be written in a very compact way by assuming that ω is a FP-ghost:

$$s\, \Delta(\omega, A) = 0 \quad . \tag{4.32}$$

In fact the BRS operator is a linear operator and therefore the equations (4.26) imply

$$s\,\Delta(\omega, A) = -\tfrac{1}{2}\, \Delta\big([\omega,\omega], A\big) + W(\omega)\,\Delta(\omega, A) \;;$$

hence (4.32) is equivalent to

$$2\, W(\omega)\, \Delta(\omega, A) - \Delta\big([\omega,\omega], A\big) = 0$$

which is precisely the WZ-condition (4.31) written in terms of an anticommuting parameter ω.

Therefore the <u>BRS operator s can be interpreted as the coboundary operator of the cohomology algebra $H^*_{loc}(\text{Lie } \mathcal{G}, \Gamma_{loc}(A))$</u> while ω can be viewed as its <u>generator</u>[94]. From this view-point the <u>integrated anomaly</u> $\Delta(\omega,A) = \int_M \alpha(\omega,A)$ corresponds to a solution of equation (4.32) which cannot be written as the s-variation of a local functional $\int_M q_4^0(A)$; indeed

$$s\int_M q_4^0(A) = W(\omega)\int_M q_4^0(A)$$

and the 4-form $q_4^0(A)$ can always be added to or substracted from the Lagrangian density.

It is convenient to reformulate this definition directly in terms of $\alpha(\omega,A)$: if M_4 is the usual Minkowski space or a 4-dimensional manifold without boundary, the equation

$$0 = s\,\Delta(\omega,A) = \int_{M_4} s\,\alpha(\omega,A)$$

is equivalent to

$$s\,\alpha(\omega,A) = -dQ_3^2 \quad \begin{array}{l}\leftarrow \text{ghost number} \\ \leftarrow \text{form degree}\end{array} \tag{4.33}$$

where Q_3^2 is a 3-form with ghost number 2. Thus an <u>anomaly</u> $\alpha(\omega,A)$ is a solution of the WZ consistency condition (4.33) which cannot be written as the s-variation of a 4-form $q_4^0(A)$. Since $sd = -ds$ these solutions are only determined up to terms of the form d(.) and <u>the determination of all possible anomalous terms amounts to an evaluation of the "s-cohomology modulo d(.)-terms"</u>[19b),43-44),138-139].

To conclude we mention a useful equivalent expression for the coboundary operator (4.30a) of the Lie algebra cohomology. This formula which is due to J.-L. Koszul[185),63] and which is nicknamed BRST-operator equation in the Physics literature[173] is discussed in Appendix 5.

(viii) <u>Algebraic determination of anomalies</u>

Although the occurence of anomalies is a typical quantum phenomenon, their determination as non trivial solutions of the WZ-consistency conditions represents a classical cohomological problem[*]. In many cases some solutions of this problem can be found in a straightforward way by Stora's descent equation method[134),135]. This purely algebraic procedure is based on a clever combination of the Chern-Weil transgression formula and a special form of the BRS equations known under the name

---

[*] An evaluation of the quantum corrections is only required for fixing the possibly vanishing coefficient of the anomalous term.

of "Russian formula". In the following it will be described in its simplest form which allows to reproduce the Adler-Bardeen (or chiral) anomaly of ordinary 4-dimensional YM-theories. For different generalizations of the transgression formula and of the descent equations we refer to 94),131) and to the subsequent chapters.

Let $J_3 : \mathfrak{g} \times \mathfrak{g} \times \mathfrak{g} \to \mathbb{R}$ be a third rank, symmetric and adG-invariant tensor on $\mathfrak{g} \equiv$ Lie G. (For compact matrix groups one considers the symmetrized trace,

$$J_3(A_1, A_2, A_3) = \mathrm{Str}(A_1, A_2, A_3) \equiv \frac{1}{3!} \sum_{\substack{\text{permut.}\\ \pi}} \mathrm{Tr}\left(A_{\pi(1)} A_{\pi(2)} A_{\pi(3)}\right) \quad (4.34))$$

Any such tensor satisfies the <u>Chern-Weil transgression formula</u> (ref. 85 ch. XII, ref. 35 appendix)

$$\begin{aligned} J_3(F, F, F) &= d\left[3 \int_0^1 dt \; J_3(A, F_t, F_t)\right] \\ &=: d\, Q_5^0(A, F) \quad . \end{aligned} \tag{4.35a}$$

Here A is a $\mathfrak{g}$-valued connection on the base space M and

$$F = dA + \frac{1}{2}[A, A]$$

$$F_t = dA_t + \frac{1}{2}[A_t, A_t] \quad , \quad A_t = tA \quad , \quad t \in [0, 1] \quad .$$

The differential forms involved in equation (4.35) are naturally defined for the 3-tensor (4.34). In the general case these expressions are to be understood as follows : given three $\mathfrak{g}$-valued forms $\alpha_1, \alpha_2, \alpha_3$ on M of respective degrees p_1, p_2, p_3, we define the $(p_1 + p_2 + p_3)$-form $J_3(\alpha_1, \alpha_2, \alpha_3)$ on M by

$$\left(J_3(\alpha_1, \alpha_2, \alpha_3)\right)_x (X_1, \dots, X_{p_1+p_2+p_3}) = \sum_\pi (-1)^\pi J_3\Big(\alpha_{1_x}(X_{\pi(1)}, \dots, X_{\pi(p_1)}), \dots, \\ \alpha_{3_x}(X_{\pi(p_1+p_2+1)}, \dots, X_{\pi(p_1+p_2+p_3)})\Big) .$$

Here $X_1, \dots, X_{p_1+p_2+p_3} \in T_x M$ and the summation is taken over all permutations π of $(1, \dots, p_1+p_2+p_3)$.

Although the 6-form $J_3(F,F,F)$ vanishes identically on a four dimensional space M, the transgression formula can be viewed and used as an algebraic device to derive well-defined and non trivial expressions on M.

We now come to the second ingredient of the method. As you may directly verify, the BRS equations for A and ω,

$$\mathfrak{s} A = -\left(d\omega + [A,\omega]\right)$$

$$\mathfrak{s}\,\omega = -\frac{1}{2}[\omega,\omega]$$

are equivalent to the so-called Russian formula (or horizontality condition[193,14]) for the curvature $F = dA + \frac{1}{2}[A,A]$ of A :

$$\mathcal{F} = F \tag{4.36}$$

where

$$\mathcal{F} \equiv \Delta \mathcal{A} + \frac{1}{2}[\mathcal{A},\mathcal{A}] \,, \quad \Delta \equiv d + \mathfrak{s} \,, \quad \mathcal{A} \equiv A + \omega \,.$$

Indeed, by expanding (4.36) in the ghost number, we obtain at ghost number 0, 1 and 2, respectively, the definition of the field strength F of A, the BRS transformation of the classical variable A and the BRS variation of the ghost parameter ω.

We observe that the Russian formula takes a slightly different form, if one considers gauge transformations and diffeomorphisms as in general relativity or in SYM-theories, see eq.(4.67).

Next we simultaneously apply the transgression and the Russian formula to obtain the equation

$$\mathfrak{J}_3(F,F,F) = \mathfrak{J}_3(\mathcal{F},\mathcal{F},\mathcal{F}) = (d+\mathfrak{s})\Big[3\int_0^1 dt\; \mathfrak{J}_3(\mathcal{A},\mathcal{F}_t,\mathcal{F}_t)\Big]$$

$$=: (d+\mathfrak{s})\; Q_5(\mathcal{A}) \tag{4.35b}$$

where

$$\mathcal{F}_t = (d+\mathfrak{s})\mathcal{A}_t + \frac{1}{2}[\mathcal{A}_t,\mathcal{A}_t] \,, \qquad \mathcal{A}_t = t\,\mathcal{A} \,.$$

By expanding the 5-form $Q_5(\mathcal{A} = A+\omega)$ in powers of ω,

$$Q_5(A+\omega) = Q_5^0(A) + Q_4^1(A,\omega) + Q_3^2(A,\omega) + Q_2^3(A,\omega) + Q_1^4(A,\omega) + Q_0^5(\omega) \tag{4.37}$$

(where the lower index denotes the form degree and the upper index the power of ω) we arrive at the descent equations by matching the corresponding powers of ω :

$$\mathfrak{J}_3(F,F,F) = (d+\mathfrak{s})\left(Q_5^0 + Q_4^1 + Q_3^2 + Q_2^3 + Q_1^4 + Q_0^5\right)$$

$$= dQ_5^0 + \left(\mathfrak{s}Q_5^0 + dQ_4^1\right) + \left(\mathfrak{s}Q_4^1 + dQ_3^2\right) + \ldots + \mathfrak{s}Q_0^5$$

hence

$$
\begin{aligned}
J_3(F,F,F) &= dQ_5^0 \\
s\,Q_5^0 + dQ_4^1 &= 0 \\
s\,Q_4^1 + dQ_3^2 &= 0 \\
&\vdots \\
s\,Q_0^5 &= 0
\end{aligned}
\tag{4.38}
$$

Comparison of the third equation with the condition (4.33) shows that the 4-form $Q_4^1(A,\omega)$ occuring in the expansion (4.37) solves the WZ-consistency condition and thus represents a candidate for the anomaly $\mathfrak{a}(A,\omega)$.

For the 3-tensor (4.34) one obtains the following explicit expression for Q_4^1:

$$Q_4^1(A,\omega) = \mathrm{Tr}\left[\omega\, d\left(A dA + \tfrac{1}{2}A^3\right)\right] \tag{4.39}$$

This is the unique anomaly occuring in 4-dimensional YM-theories, up to trivial terms of the form $s\alpha + d\beta$ and up to a numerical factor which can be evaluated by computing a convergent Feynman diagram ; the latter is known to be non vanishing only in the case where chiral fermions are coupled to A[4),5),13),19),134)].

For an abelian gauge group we may choose $J_3(F,F,F) = F^3$ with the result that

$$Q_4^1(A,\omega) = \omega(dA)(dA) = \omega F F$$

This quantity can be rewritten as

$$
\begin{aligned}
\omega\left(F_{\mu\nu}\, dx^\mu \wedge dx^\nu\right)\wedge\left(F_{\rho\sigma}\, dx^\rho \wedge dx^\sigma\right) &= \omega\, F_{\mu\nu} F_{\rho\sigma}\left(dx^\mu \wedge dx^\nu \wedge dx^\rho \wedge dx^\sigma\right) \\
&= \omega\left(\varepsilon^{\mu\nu\rho\sigma} F_{\mu\nu} F_{\rho\sigma}\right) dx^0 \wedge dx^1 \wedge dx^2 \wedge dx^3
\end{aligned}
\tag{4.40}
$$

where we recognize the familiar expression for the axial U(1) anomaly[175),3),80),81)].

A few comments concerning the different contributions to the expansion (4.37) are in order. As shown by M. Atiyah and I.M. Singer[12)] the 4-forms $Q_4^{2n-1}(A,\omega)$ in which the parameter ω occurs in $2n-1$ times $(n \geq 1)$ correspond to obstructions to the existence of a covariant Dirac propagator in an external YM-field ; yet it is still unclear, if the higher obstructions $(Q_4^3, Q_4^5, \ldots)$ have any physical significance.

On the other hand L. Fadeev has recently demonstrated[48)] that a cocycle[*)] of

---

*) i.e. a cocycle in the cohomology of s modulo d (see eqs (4.38) and previous section).

the form Q_3^2 occurs as a Schwinger term in the Hamiltonian approach to gauge theories in 3-spatial dimensions $(t = 0)$. Most higher cocycles can be given a field theoretic interpretation within the framework of local BRS current algebra[17)].

To conclude we note that the chiral anomaly is related to the index of the family of Dirac operators $A \to \not{D}(A)$ and that the Atiyah-Singer index theorem provides an alternative and powerful tool for evaluating the anomalous terms. (This was realized in the mid-seventies by N.K. Nielsen, H. Römer and B. Schroer[103),176)]; for extensive reviews of the recent work we refer to 6),124),125)). This approach automatically gives the correct normalization of the anomaly[164)]. However it relies on the ordinary de Rham cohomology and still lacks a proper incorporation of the concept of locality[125),136)].

IV.2 BRS algebras and anomalies in SYM-theories

(i) Superfield algebras in the chiral representation

As pointed out in part II the chiral basis is best suited for the quantization of SYM-theories. The basic variable is the real gauge superfield V transforming with a chiral parameter $\Lambda = \Lambda^r T_r$ under supergauge transformations

$$e^{V'} = e^{-i\Lambda^\dagger} e^V e^{i\Lambda} = \left(1 - i\Lambda^\dagger + \dots\right) e^V \left(1\!\!1 + i\Lambda + \dots\right)$$

$$= e^V - \left(i\Lambda^\dagger\right) e^V + e^V \left(i\Lambda\right) + \text{ terms quadratic in } \Lambda, \Lambda^\dagger .$$

For the infinitesimal transformations one has

$$\delta_\Lambda e^V = e^V (i\Lambda) - (i\Lambda^\dagger) e^V .$$

Turning $i\Lambda \equiv c$ and $i\Lambda^+ \equiv \bar{c}$ into Fadeev-Popov ghosts with ghost numbers $+1$ and -1, respectively, we are led to the <u>differential algebra</u>*)

$$\begin{aligned} s\, e^V &= - e^V c + \bar{c}\, e^V \\ s\, c &= -\tfrac{1}{2}[c, c] = - c c \\ s\, \bar{c} &= -\tfrac{1}{2}[\bar{c}, \bar{c}] = -\bar{c}\bar{c} . \end{aligned} \tag{4.41}$$

These relations represent an essential part of the algebra considered in ref. 62)

---

*) According to our superspace conventions the antiderivation s satisfies the graded Leibniz rule (4.55).

(see also ref. 66)) for determining a superfield expression of the susy chiral anomaly. Similar expressions have been obtained by analogous and by different methods[26),56),68),97),104),110),113)], but a direct comparison of these results is quite difficult, since the susy chiral anomaly has necessarily a non-polynomial character[52)] and as any anomaly it is only determined up to s and d-terms (see however ref. 184)). For the final results which are fairly complicated and not very illuminating we refer to the original articles.

Concerning the possible supersymmetry anomalies we have the following fundamental result due to O. Piguet, M. Schweda and K. Sibold[112)] : assuming translation invariance of the theory all supersymmetry anomalies on superspace are cohomologically trivial.

After these short comments on the anomalies in the superfield formalism, we continue our presentation of the superspace differential algebras. To discuss the quantization and renormalization of the theory one considers the s-variation of the basic superfield V :

$$
\begin{aligned}
s e^{V} &= sV + \frac{1}{2} V(sV) + \frac{1}{2}(sV)V + \ldots \\
= -e^{V} c + \bar{c} e^{V} &= -(\mathbb{1} + V + \ldots)c + \bar{c}(\mathbb{1} + V + \ldots) \\
&= -c + \bar{c} - Vc + \bar{c} V + \ldots
\end{aligned}
$$

hence

$$
\begin{aligned}
sV &= -c + \bar{c} - \frac{1}{2}(Vc - \bar{c}V + V\bar{c} - cV) + \ldots \\
&= -c + \bar{c} - \frac{1}{2}[V, c + \bar{c}] + \ldots \\
&= -c + \bar{c} - \sum_{n \geq 1} h_n \underset{(1)}{[V,} \underset{(2)}{[V,} \ldots, \underset{(n)}{[V,} c - (-1)^n \bar{c}], \ldots] \\
&\equiv Q(V, c)
\end{aligned}
$$

After introducing the chiral anti-ghost superfield c_- and the chiral auxiliary superfield $B = B^r T_r$ we have the following superspace generalization of the BRS differential algebra (4.13):

$$s V = Q(V, c)$$

$$s c = -\frac{1}{2}[c, c] \qquad s \bar{c} = -\frac{1}{2}[\bar{c}, \bar{c}]$$

$$s c_- = B \qquad s \bar{c}_- = \bar{B}$$

$$s B = 0 \qquad s \bar{B} = 0 \tag{4.42}$$

This algebra represents the starting point for the BRS quantization of SYM-theories[114),57)].

In this approach one usually chooses a <u>supergauge fixing term</u>

$$-\frac{1}{\lambda} \mathrm{Tr} \int d^4x \int d^4\theta \, (D^2V)(\bar{D}^2V)$$

which contains a component field contribution $(\partial_m v^m)^2$ and thereby generalizes the Lorentz gauge condition to superspace. This supersymmetric ("covariant") gauge treatment allows to apply the powerful supergraph techniques, but it considerably complicates all the general proofs of renormalizability and gauge independence (see 114),90)). These difficulties which have been overcome in recent years[67),114)] can be traced back to the fact that the superfield V and in particular its lowest component C are dimensionless ; consequently there is a propagator of the type $1/k^4$ which leads to spurious infrared singularities in the off-shell Green functions.

Several approaches have recently been considered to bypass these problems. We mention the work of L.F. Abbott et al.[1)] who used non-local gauge fixing terms and of W. Kummer et al.[90)] who studied the quantization in non-supersymmetric WZ-type gauges while trying to preserve as much as possible of the supergraph techniques.

Finally we note that the field theory approaches which are based on functional integrals can only be given a well-defined sense in Euclidean space[174)]. Though the formulation of <u>supersymmetric Euclidean field theories</u> is not a trivial matter, since "chirality" of fermion fields and Majorana spinors do not represent Euclidean covariant concepts. For some recent progress in this direction we refer to 95) and to the references given therein.

(ii) BRS algebra and anomalies in the WZ-gauge

In order to obtain simple component field expressions for the anomalies in SYM-theories, it seems natural to consider the WZ-gauge. This gauge represents a possible choice for the N=1, d=4 theory, but in higher dimensions and for certain 4-dimensional extended susy theories without a complete superspace formulation there is no alternative. In the following we will set up the Lie algebra which yields the WZ-integrability conditions for the corresponding anomalies and we will introduce the associated differential algebra.

To start with we consider the variations of the classical fields v_m, λ_α and D under infinitesimal gauge transformations, susy transformations and translations*). These transformations are described, respectively, by a real Lie algebra valued parameter $v(x)$, an anticommuting variable ζ and a real four-vector $a = (a^m)$:

$$\begin{aligned}
&\delta_a v_m = -(a\cdot\partial) v_m \;, && \delta_v v_m = \mathcal{D}_m v \;, && \delta_\zeta v_m = i\left[\zeta\sigma_m\bar\lambda + \bar\zeta\bar\sigma_m\lambda\right] \\
&\delta_a \lambda_\alpha = -(a\cdot\partial)\lambda_\alpha \;, && \delta_v \lambda_\alpha = -[v,\lambda_\alpha] \;, && \delta_\zeta \lambda_\alpha = (\sigma^{mn}\zeta)_\alpha v_{mn} + i\zeta_\alpha D \\
&\delta_a D = -(a\cdot\partial) D \;, && \delta_v D = -[v,D] \;, && \delta_\zeta D = -\zeta\sigma^m(\mathcal{D}_m\bar\lambda) + \bar\zeta\bar\sigma^m(\mathcal{D}_m\lambda)
\end{aligned} \tag{4.43}$$

The induced functional derivatives

$$W(a) = \mathrm{Tr}\int d^4x \left\{ (\delta_a v_m)\frac{\delta}{\delta v_m} + (\delta_a\lambda_\alpha)\frac{\delta}{\delta\lambda_\alpha} + (\delta_a\bar\lambda^{\dot\alpha})\frac{\delta}{\delta\bar\lambda^{\dot\alpha}} + (\delta_a D)\frac{\delta}{\delta D}\right\} \tag{4.44}$$

(similarly for $W(\zeta)$, $W(v)$) satisfy the following commutation relations ($a_{12}^m \equiv 2i(\zeta_1\sigma^m\bar\zeta_2 - \zeta_2\sigma^m\bar\zeta_1)$):

$$\left[W(a), W(a')\right] = \left[W(a), W(\zeta)\right] = 0 \tag{4.45a}$$

$$\left[W(\zeta_1), W(\zeta_2)\right] = W(a_{12}) + W(a_{12}^m v_m) \tag{4.45b}$$

$$\left[W(a), W(v)\right] = W(-a^m\partial_m v) \tag{4.45c}$$

$$\left[W(\zeta), W(v)\right] = 0 \tag{4.45d}$$

$$\left[W(v), W(v')\right] = W([v,v']) \; . \tag{4.45e}$$

As first order differential operators $W(a)$, $W(\zeta)$ and $W(v)$ automatically satisfy the Jacobi identity so that the relations (4.43-45) actually define a Lie algebra.

*) The translations are required for the closure of the susy algebra.

Note however that a triple commutator involving the v_m-dependent term $[W(\zeta_1),W(\zeta_2)]$ is not directly determined by the relations (4.45), rather it has to be evaluated explicitly since the functional derivatives W(.) contain variations with respect to the field v_m.

Equation (4.45b) reflects the well-known fact[152),40),128)] that the algebra of the "susy transformations in the WZ-gauge" is that of the constrained gauge covariant derivatives (ch. III.2).

Next we consider the WZ-consistency conditions induced by the commutation relations (4.45). Note that the vertex functional S_{quant} is invariant under translations, if the theory is properly defined in terms of tempered distributions[22)]:

$$W(a)\, S_{quant} = 0 \; .$$

Denoting the gauge anomaly by G(v) and the "susy anomaly" by S(ζ), we have

$$W(v)\, S_{quant} = G(v;\, v_m, \lambda, D)$$

$$W(\zeta)\, S_{quant} = S(\zeta;\, v_m, \lambda, D) \; .$$

Thereby (4.45) implies

$$W(v)\, G(v') - W(v')\, G(v) - G([v, v']) = 0 \qquad (4.46a)$$

$$W(v)\, S(\zeta) - W(\zeta)\, G(v) = 0 \qquad (4.46b)$$

$$W(\zeta_1)\, S(\zeta_2) - W(\zeta_2)\, S(\zeta_1) - G(a_{12}^m v_m) = 0 \; . \qquad (4.46c)$$

As these equations are inhomogeneous in the susy anomaly, a non zero gauge anomaly induces a non vanishing susy anomaly despite the fact that such terms are cohomologically trivial at the superfield level. An explicit expression for S(ζ) was determined by H. Itoyama, V.P. Nair and H.C. Ren[78)*)] and by E. Guadagnini and M. Mintchev[69)] (for d=2,10 see 76),32))

$$S(\zeta) = \int d^4x \; \mathrm{str}\left\{ (\delta_\zeta A)\left[A(dA) + (dA)A + \frac{3}{2} A^3 + \frac{i}{4} \bar{\lambda}\, \gamma^{(3)} \lambda \right] \right\} \qquad (4.47)$$

where

---

*) A correction of the initial result is presented in the second paper of ref. 78).

$$A = A_m dx^m \quad , \quad \lambda = \begin{pmatrix} \lambda_\alpha \\ \bar{\lambda}^{\dot{\alpha}} \end{pmatrix}$$

$$\bar{\lambda}\, \gamma^{(3)} \lambda = \frac{1}{6} (\bar{\lambda}\, \gamma_{[l}\, \gamma_m\, \gamma_{n]}\, \lambda)\, dx^l \wedge dx^m \wedge dx^n \quad .$$

The result which was verified to be cohomologically non-trivial does not depend on the auxiliary field D and consists of two distinct terms ; the second contribution which only depends on λ and $\bar{\lambda}$ is gauge invariant. The quantity (4.47) coincides with the expression that follows from some of the superfield formulae by restriction to the WZ-gauge[149),27)].

Along the lines of section IV.1 (iv) we now introduce the differential algebra associated to the given Lie algebra by turning v, ζ and a^m into FP-ghosts. Thus v and a^m (resp. ζ) become anticommuting (resp. commuting) variables ; the formal details of this grading are to be discussed in the next chapter. On the classical fields v_m, λ_α and D the s-variation corresponds to the sum of the infinitesimal variations (4.43)[*)]:

$$s v_m = \mathcal{D}_m v - (a\cdot\partial) v_m - i\left[\zeta \sigma_m \bar{\lambda} + \bar{\zeta}\bar{\sigma}_m \lambda\right] \tag{4.48a}$$

$$s \lambda_\alpha = \frac{i}{2}\left[v, \lambda_\alpha\right] - (a\cdot\partial)\lambda_\alpha - \left[(\sigma^{mn}\zeta)_\alpha v_{mn} + i\zeta_\alpha D\right] \tag{4.48b}$$

$$s D = \frac{i}{2}\left[v, D\right] - (a\cdot\partial) D + \left[\zeta\sigma^m(\mathcal{D}_m \bar{\lambda}) - \bar{\zeta}\bar{\sigma}^m(\mathcal{D}_m \lambda)\right] \tag{4.48c}$$

When turning ζ_1 and ζ_2 into FP-ghosts, the translation parameter

$$a^m_{12} = 2i(\zeta_1\sigma^m\bar{\zeta}_2 - \zeta_2\sigma^m\bar{\zeta}_1) = -2i(\bar{\zeta}_2\bar{\sigma}^m\zeta_1 + \zeta_2\sigma^m\bar{\zeta}_1)$$

becomes

$$a^m = -2i(\bar{\zeta}\bar{\sigma}^m\zeta + \zeta\sigma^m\bar{\zeta})$$

$$= -2i(\zeta\sigma^m\bar{\zeta} + \zeta\sigma^m\bar{\zeta}) = -4i\,(\zeta\sigma^m\bar{\zeta}) .$$

In view of the equations (4.19),(4.45a,b,d) we define

$$s\, a^m = -\frac{1}{2}\left[-4i(\zeta\sigma^m\bar{\zeta})\right] = 2i\,(\zeta\sigma^m\bar{\zeta}) \tag{4.48d}$$

$$s\, \zeta^\mu = 0 = s\, \bar{\zeta}_{\dot{\mu}} \tag{4.48e}$$

---

*) The numerical factor in the v-dependent contributions is irrelevant and was introduced to have coincidence with the results of the geometric approach considered hereafter.

Equations (4.45e,c) suggest that the s-variation of v contains the usual Maurer-Cartan term and an ordinary translation contribution ; but the "susy algebra" (4.45a,b) only closes modulo a gauge transformation with a field-dependent parameter and therefore the latter also appears in sv :

$$s v = -\frac{i}{4}\left[v, v\right] - (a\cdot\partial)v - 2i(\zeta\sigma^m\bar{\zeta})v_m \quad . \tag{4.48f}$$

The so-defined s-operations are nilpotent and define the "BRS algebra in the WZ-gauge". They first appeared in Zumino's Argonne lectures[165)].

Although several authors considered the quantization of component fields in the WZ-gauge[29),78)], none of these approaches is based on the given differential algebra. The reasons for avoiding this formulation are the potential infrared divergences which are related to the presence of a constant external field ζ of dimension $-1/2$, i.e. to a superrenormalizable coupling. Likely these difficulties can be overcome by assuming ζ to be space-time dependent and studying SYM-theories coupled to external supergravity ; subsequently the proper quantum theory should be recovered in the "adiabatic limit" of a constant field ζ.

IV.3 Geometric approach to the "BRS algebra in the WZ-gauge"*)

The fact that the Lie algebra and the differential algebra introduced in the last section depend on the dynamical fields of the theory might be physically appealing, but it is mathematically obscure[136)]; in particular the associated anomalies are not properly defined from the mathematical point of view. In the following we will try to gain a better geometric understanding of these algebras and we analyze whether or not they can be given a field independent form by an appropriate field dependent change of variables.

First we will discuss these questions at the superfield level while following closely the authors of ref. 91) who studied the analogous problem for chiral fermions interacting with an external gravitational field. (Though in our rigid superspace approach there is no need for the background fields introduced in ref.91) to account for a possible non-trivial topology of the base manifold.) Thus the (BRS-) differential algebra will be defined on rigid superspace in terms of the superconnection $\mathbb{A}$ and the associated curvature $\mathbb{F}$. By implementing the general solution of the constrained Bianchi identities and by projecting the superfield equations in the gauge vector representation we reproduce Zumino's component field algebra as well as the corresponding N=2 algebra. This geometric approach is

*) This chapter is based on a recent article by the author[60)].

based on the general methods developed for supergravity by L. Baulieu, M. Bellon and R. Grimm[18].

To conclude it will be argued that, without introducing additional fields, the N=1 component field algebra can<u>not</u> be given field <u>in</u>dependent form. This result can be traced back to the fact that the susy transformations do not act on the ordinary gauge transformations in the WZ-gauge algebra. Consequently the mathematical characterization of the "susy anomalies in the WZ-gauge" remains obscure.

(i) General framework

The following rigid superspace considerations are quite general and there is no need to specify the dimension d of space-time or the particular representation (gauge vector or chiral representation) for the superconnection $\mathbb{A} = E^A \mathbb{A}_A$ and the matter superfield Φ ; also the superspace indices A, B,... may include extended susy algebra indices, $A = (a, \alpha i, \dot{\alpha} j)$ with $i,j \in \{1,\dots,N\}$ (see part V).

The infinitesimal supergauge resp. susy transformations of the basic superfields are given by

$$\begin{aligned} \delta_{\mathfrak{X}} \Phi &= -\mathfrak{X} \Phi & \text{resp.} \qquad \delta_\xi \Phi &= -L_\xi \Phi \\ \delta_{\mathfrak{X}} \mathbb{A} &= -\mathcal{D}\mathfrak{X} & \delta_\xi \mathbb{A} &= -L_\xi \mathbb{A} \\ \delta_{\mathfrak{X}} E^A &= 0 & \delta_\xi E^A &= -L_\xi E^A = 0 \, . \end{aligned} \tag{4.49}$$

Here $\mathfrak{X}$ represents a Lie algebra valued superfield and ξ is the supervector field (1.5) (including ordinary translations which are required for the closure of the final component field algebra).

The induced variation of a functional $F[E^A, \mathbb{A}, \Phi]$ is described by the functional derivatives

$$W(\mathfrak{X}) = \int d^8 z \left[(\delta_{\mathfrak{X}} \mathbb{A}) \frac{\delta}{\delta \mathbb{A}} + (\delta_{\mathfrak{X}} \Phi) \frac{\delta}{\delta \Phi} \right] \tag{4.50}$$

(similarly for $W(\xi)$) whose commutation relations are the following:

$$[W(\mathfrak{X}), W(\mathfrak{X}')] = W([\mathfrak{X}, \mathfrak{X}']) \tag{4.51a}$$

$$[W(\xi), W(\mathfrak{X})] = W(L_\xi \mathfrak{X}) = W(i_\xi d\mathfrak{X}) \tag{4.51b}$$

$$[W(\xi), W(\xi')] = W([\xi, \xi']) \tag{4.51c}$$

Here $[\mathfrak{X},\mathfrak{X}']$ represents the commutator of Lie algebra valued superfields and $[\xi,\xi']$ is the Lie bracket of supervector fields (eq.(1.7)).

We now briefly discuss the structure of the algebra (4.49-51). Together with the relations (4.51) the vector space $\varepsilon \equiv \{W(\mathfrak{X}),W(\xi)\}$ of infinitesimal variations defines a Lie algebra. The equations (4.51a,b) show that the commutator $[W(\mathfrak{X}),W(.)]$ is again a supergauge variation and therefore the vector subspace $\{W(\mathfrak{X})\}$ of ε is an ideal of the Lie algebra ε. Hence we can consider the quotient-space $\varepsilon/\{W(\mathfrak{X})\}$, i.e. the set of all equivalence classes $\widehat{W(.)} \equiv \{W(.) + W(\mathfrak{X})\}$. This space again represents a Lie algebra with a commutator defined by

$$\left[\widehat{W_1(\cdot)}, \widehat{W_2(\times)}\right] := \widehat{\left[W_1(\cdot), W_2(\times)\right]}$$

$\varepsilon/\{W(\mathfrak{X})\}$ is parametrized by the variations $W(\xi)$ and is isomorphic to the Lie algebra $\{\xi\}$ of supervector fields.

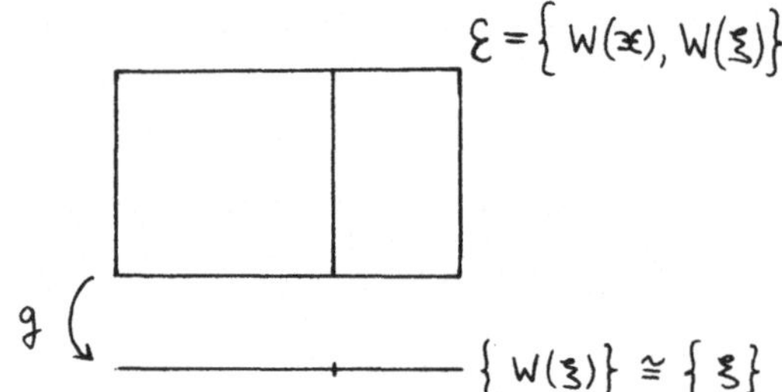

We note that the sequence

$$\{0\} \overset{j}{\hookrightarrow} \{W(\mathfrak{X})\} \overset{f}{\hookrightarrow} \mathcal{E} \equiv \{W(\mathfrak{X}), W(\xi)\} \overset{g}{\twoheadrightarrow} \underset{\cong\{\xi\}}{\{W(\xi)\}} \overset{h}{\twoheadrightarrow} \{0\}$$

(where j,f denote the inclusion and g,h the projection maps) is a short exact sequence of Lie algebras: j,f,g,h are Lie algebramorphisms and

$$\operatorname{Im} j = \{0\} = \operatorname{Ker} f$$

$$\operatorname{Im} f = \{W(\mathfrak{X})\} = \operatorname{Ker} g$$

$$\operatorname{Im} g = \{\xi\} = \operatorname{Ker} h \quad .$$

Thus the Lie algebra ε is an extension[70)] of $\{\xi\} \cong \varepsilon/\{W(\mathfrak{X})\}$ by $\{W(\mathfrak{X})\}$.

Actually ε is a very particular extension, because it has a semi-direct sum structure:

$$\mathcal{E} = \{W(\xi)\} \oplus \{W(\mathfrak{X})\} \tag{4.52}$$

("The translations and susy transformations act on the supergauge transformations".)

Before turning χ and ξ into Fadeev-Popov ghosts we observe that there is already a double grading in the theory, since all superfields and forms Q_A are characterized by their differential form degree and by their statistics degree ("Grassmann parity"). Although this double grading is equivalent to a simple grading by the sum of the two degrees, it is more convenient to work with the bi-graded quantities and to assume that the statistics degree has only a book-keeping role (it is not affected by the exterior differential d) (section I.5 (i)).

We now turn χ and ξ into FP ghosts by assigning a ghost number one to the components of these superfields. As in the bosonic case we add the ghost number to the form degree while the statistics degree is assumed to keep its passive role (it is not affected by the BRS operator s). Accordingly the graded commutator of two $\mathfrak{g}$-valued forms P_A, Q_B of respective total degrees p,q reads

$$[P_A, Q_B] = P_A Q_B - (-)^{ab+pq} Q_B P_A \tag{4.53}$$

and the graded Lie bracket of two supervector fields ξ, η with ghost numbers deg ξ, deg η is given by

$$[\xi, \eta] = \left[\xi^N(\partial_N \xi^M) - (-)^{(\deg \xi)(\deg \eta)} \eta^N(\partial_N \xi^M)\right] \partial_M \quad . \tag{4.54}$$

On this graded algebra the s-operator acts as an antiderivation of degree 1 which anticommutes with the differential d (when acting on forms), $sd + ds = 0$. With our conventions for the super-calculus the anti-Leibniz rule for s (and d) reads

$$s(PQ) = P(sQ) + (-)^q (sP)Q \quad \text{for} \quad \deg Q = q \; . \tag{4.55}$$

As in the bosonic case we have the (anti-) commutation relations

$$[s, i_\xi] = i_{s\xi} \quad , \quad \{s, L_\xi\} = L_{s\xi} \tag{4.56}$$

where i_ξ represents a derivation annihilating 0-forms and where $L_\xi \equiv i_\xi d - d i_\xi$ is an antiderivation of degree 1.

(ii) Superfield differential algebra

The differential algebra associated to the Lie algebra ε given by (4.49-51) reads*)

$$s E^A = 0 \tag{4.57a}$$

$$s \Phi = \mathcal{X} \Phi - L_\xi \Phi \tag{4.57b}$$

$$s \mathbb{A} = -\mathcal{D}\mathcal{X} - L_\xi \mathbb{A} \tag{4.57c}$$

$$s \mathcal{X} = -\frac{1}{2}[\mathcal{X}, \mathcal{X}] - L_\xi \mathcal{X} \tag{4.57d}$$

$$s \xi = -\frac{1}{2}[\xi, \xi] \tag{4.57e}$$

where $\mathcal{D}\mathcal{X} = d\mathcal{X} + [\mathbb{A}, \mathcal{X}]$ (see def.(2.20)). From these relations one directly deduces the transformation law of the curvature 2-form $\mathbb{F} = d\mathbb{A} + \mathbb{A}\mathbb{A}$ by using the basic properties of the differential s :

$$s \mathbb{F} = -[\mathbb{F}, \mathcal{X}] - L_\xi \mathbb{F} \tag{4.57f}$$

The so-defined operator s is nilpotent and thus can be considered as the coboundary operator for the local cohomology of the Lie algebra ε with values in the vector space $\Gamma_{loc}(E^A, \mathbb{A})$ (ch. IV.1).

In the gauge real representation which will be chosen for projecting the theory to space-time, the susy transformations are described by the supergauge covariant Lie derivative $\mathcal{L}_\xi$ (eq.(3.4)). Therefore we consider the change of generators $\tilde{\mathcal{X}} := \mathcal{X} - i_\xi \mathbb{A}$ in the differential algebra (4.57) : from $d\Phi = \mathcal{D}\Phi + \mathbb{A}\Phi$ it then follows that

$$\begin{aligned} s \Phi &= \mathcal{X}\Phi - i_\xi d\Phi \\ &= \mathcal{X}\Phi - i_\xi \mathcal{D}\Phi - (i_\xi \mathbb{A})\Phi \\ &= \tilde{\mathcal{X}}\Phi - i_\xi \mathcal{D}\Phi \quad . \end{aligned} \tag{4.58a}$$

Similarly, by using $d\mathbb{A} = \mathbb{F} - \mathbb{A}\mathbb{A}$, the Bianchi identity $\mathcal{D}\mathbb{F} \equiv d\mathbb{F} - [\mathbb{A}, \mathbb{F}] = 0$ and $i_{[\xi,\eta]} = [L_\xi, i_\eta]$ the remaining equations can be rewritten as follows:

---

*) As in the bosonic case the relative signs of the different contributions have to be defined in such a way that $s^2 = 0$.

$$s\, E^A = 0 \tag{4.58b}$$

$$s\, \mathbb{A} = -\mathcal{D}\tilde{\mathcal{X}} - i_\xi \mathbb{F} \tag{4.58c}$$

$$s\, \mathbb{F} = -[\mathbb{F}, \tilde{\mathcal{X}}] + \mathcal{D} i_\xi \mathbb{F} \tag{4.58d}$$

$$s\, \tilde{\mathcal{X}} = -\frac{1}{2}[\tilde{\mathcal{X}}, \tilde{\mathcal{X}}] + \frac{1}{2} i_\xi i_\xi \mathbb{F} \tag{4.58e}$$

$$s\, \xi = -\frac{1}{2}[\xi, \xi] \tag{4.58f}$$

The s-invariance of E^A implies

$$s\, \mathbb{A} = E^A (s\, \mathbb{A}_A)$$

and with $i_\xi \mathbb{F} = E^A \xi^B \mathbb{F}_{BA}$ we thus obtain

$$s\, \mathbb{A}_A = -\mathcal{D}_A \tilde{\mathcal{X}} - \xi^B \mathbb{F}_{BA} \tag{4.59a}$$

Analogously we get

$$s\, \mathbb{F}_{BA} = -[\mathbb{F}_{BA}, \tilde{\mathcal{X}}] - \xi^C \mathcal{D}_C \mathbb{F}_{BA} \tag{4.59b}$$

by applying the Bianchi identity for $\mathbb{F}$:

$$\begin{aligned} \mathcal{D} i_\xi \mathbb{F} &= (\mathcal{D} i_\xi - i_\xi \mathcal{D})\mathbb{F} = -\mathcal{L}_\xi \mathbb{F} \\ &= -\frac{1}{2} E^A E^B (\mathcal{L}_\xi \mathbb{F}_{BA}) = -\frac{1}{2} E^A E^B (i_\xi \mathcal{D} \mathbb{F}_{BA}) \\ &= -\frac{1}{2} E^A E^B (\xi^C \mathcal{D}_C \mathbb{F}_{BA}) \end{aligned}$$

Clearly the familiar constraints of the $N = 1,2$ theories (eqs (2.27) and (5.16)) are invariant under the transformations (4.59b).

Thanks to the complete covariance of our approach the relations (4.58a) and (4.59b) remain valid for the gauge covariant derivatives of the superfields Φ and $\mathbb{F}_{BA}$ respectively (i.e. for $\mathcal{D}_A\Phi, \ldots, \mathcal{D}_C\mathbb{F}_{BA}, \ldots$).

Before discussing the projection from superspace to space-time we will make a few unrelated comments concerning the differential algebras introduced in this section.

1. If we assume that the relations (4.57) define the BRS algebra of the real basis, we can formally introduce those of the other representations by considering the BRS transformations of the prepotentials $\mathcal{U}$ and $\mathcal{V}$. For instance from the transformation law (2.52) of $\mathcal{U}$,

$$\mathcal{U}' = \Sigma^{-1}\,\mathcal{U}\,X = e^{-i\Lambda}\,\mathcal{U}\,e^{\mathcal{X}}$$

$$\simeq (\mathbb{1} - i\Lambda)\,\mathcal{U}(\mathbb{1} + \mathcal{X}) \simeq \mathcal{U} + \mathcal{U}\mathcal{X} - i\Lambda\,\mathcal{U}$$

we have the infinitesimal variation

$$\delta_{\mathcal{X},\Lambda}\,\mathcal{U} = \mathcal{U}\mathcal{X} - i\Lambda\,\mathcal{U}$$

which suggests the definition of the following (nilpotent) s-operation:

$$s\,\mathcal{U} = -\,\mathcal{U}\mathcal{X} + i\Lambda\,\mathcal{U} - L_\xi\,\mathcal{U} \tag{4.60a}$$

$$s\,\Lambda = -\frac{i}{2}[\Lambda,\Lambda] - L_\xi\,\Lambda\;. \tag{4.60b}$$

The corresponding transformation of $\mathcal{U}^{-1}$ can be derived from

$$0 = s(\mathcal{U}\mathcal{U}^{-1}) = \mathcal{U}(s\,\mathcal{U}^{-1}) + (s\,\mathcal{U})\,\mathcal{U}^{-1}$$

and it reads

$$s\,\mathcal{U}^{-1} = \mathcal{X}\,\mathcal{U}^{-1} - i\,\mathcal{U}^{-1}\Lambda - L_\xi\,\mathcal{U}^{-1}\;. \tag{4.60c}$$

Now the BRS variations of the chiral representation variables $\phi = \mathcal{U}\Phi$, $\varphi = \mathcal{U}\mathbf{A}\,\mathcal{U}^{-1} - \mathcal{U}d\,\mathcal{U}^{-1}$ and $\mathcal{F} = \mathcal{U}\mathbf{F}\,\mathcal{U}^{-1}$ immediately follows from those of Φ, $\mathbf{A}$, $\mathbf{F}$ and $\mathcal{U}, \mathcal{U}^{-1}$:

$$s\,\phi = i\Lambda\phi - L_\xi\,\phi \tag{4.61a}$$

$$s\,\varphi = -i\,\mathcal{D}^{(\varphi)}\Lambda - L_\xi\,\varphi \tag{4.61b}$$

$$s\,\mathcal{F} = -i[\mathcal{F},\Lambda] - L_\xi\,\mathcal{F}\;. \tag{4.61c}$$

Similarly the corresponding algebra of the anti-chiral basis can be deduced from the definition

$$s\mathcal{V} = -\mathcal{V}\mathcal{X} + i\bar{\Lambda}\mathcal{V} - L_\xi\mathcal{V} \tag{4.62a}$$

$$s\mathcal{V}^{-1} = \mathcal{X}\mathcal{V}^{-1} - i\mathcal{V}^{-1}\bar{\Lambda} - L_\xi\mathcal{V}^{-1} \tag{4.62b}$$

$$s\bar{\Lambda} = -\frac{i}{2}[\bar{\Lambda},\bar{\Lambda}] - L_\xi\bar{\Lambda} \tag{4.62c}$$

Both algebras are related by the s-variation of the prepotential $\mathcal{W} = \mathcal{V}\mathcal{U}^{-1}$:

$$s\mathcal{W} = s(\mathcal{V}\cdot\mathcal{U}^{-1}) = -i\mathcal{W}\Lambda + i\bar{\Lambda}\mathcal{W} - L_\xi\mathcal{W} \tag{4.63a}$$

$$s\mathcal{W}^{-1} = s(\mathcal{U}\cdot\mathcal{V}^{-1}) = i\Lambda\mathcal{W}^{-1} - i\mathcal{W}^{-1}\bar{\Lambda} - L_\xi\mathcal{W}^{-1}. \tag{4.63b}$$

Thus we recover the <u>triangular structure</u> described in chapter II for the supergauge transformations:

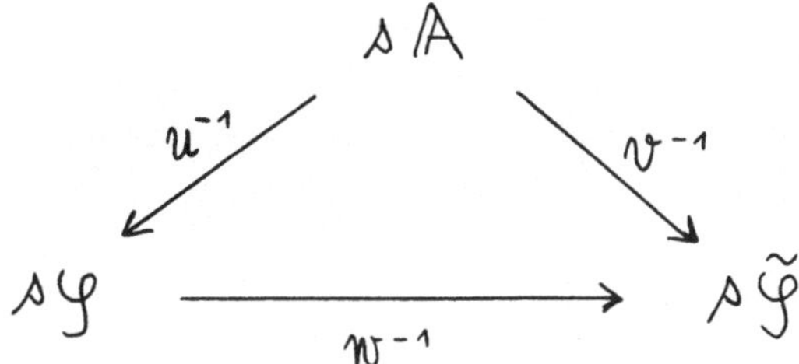

2. The BRS algebra of a gauge theory involving space-diffeomorphisms can be cast into a very compact form by the following trick[14]. Consider the differential algebra (4.58) and define

$$\hat{s} := s + L_\xi \tag{4.64a}$$

$$\hat{\mathcal{X}} := \tilde{\mathcal{X}} + i_\xi A \tag{4.64b}$$

Then

$$\hat{s}^2 = 0 \quad \Longleftrightarrow \quad s^2 = 0\ ,\quad s\xi = -\frac{1}{2}[\xi,\xi]$$

and

$$\hat{s}d + d\hat{s} = 0 \quad \Longleftrightarrow \quad sd + ds = 0 \quad (L_\xi = i_\xi d - di_\xi)\,.$$

Furthermore the s-variations (4.58) are equivalent to the following ŝ-variations:

$$\hat{s}\, E^A = 0$$

$$\hat{s}\, \Phi = \hat{\mathcal{X}}\, \Phi$$

$$\hat{s}\, \mathbb{A} = -\mathcal{D}\hat{\mathcal{X}}$$

$$\hat{s}\, \mathbb{F} = -[\mathbb{F}, \hat{\mathcal{X}}]$$

$$\hat{s}\, \hat{\mathcal{X}} = -\frac{1}{2}[\hat{\mathcal{X}}, \hat{\mathcal{X}}]$$

$$\hat{s}\, \xi = +\frac{1}{2}[\xi, \xi] \tag{4.65}$$

If we disregard the transformation of ξ, this algebra has the same form as the BRS algebra of a gauge theory without space-diffeomorphisms.

The introduction of the <u>"effective s-operation and ghost fields"</u> (4.64) is particularly useful for dealing with more complicated algebras like those occuring in supergravity[16),18)]. In the present situation it is very convenient for explicitly verifying the postulated covariance of the s-variations of Φ and $\mathbf{F}$. For instance $\hat{s}\Phi = +\hat{\mathcal{X}}\Phi$ immediately yields the (nilpotent) transformation

$$\hat{s}\,(\mathcal{D}\Phi) = -\hat{\mathcal{X}}\,(\mathcal{D}\Phi)$$

from which we deduce the relation

$$s\,(\mathcal{D}\Phi) = -\tilde{\mathcal{X}}\,(\mathcal{D}\Phi) - \mathcal{L}_\xi\,(\mathcal{D}\Phi)$$

and the postulated s-variation

$$s\,(\mathcal{D}_A\Phi) = +\tilde{\mathcal{X}}\,(\mathcal{D}_A\Phi) - \xi^B\,(\mathcal{D}_B\mathcal{D}_A\Phi)\,.$$

(Similarly for $s(\mathcal{D}_A\mathcal{D}_B\Phi)$ and $s(\mathcal{D}_C\mathbf{F}_{BA})$).

<u>3.</u> The <u>Lie algebra associated to the differential algebra (4.58)</u> is given by*)

$$[W(\tilde{\mathcal{X}}), W(\tilde{\mathcal{X}}')] = W([\tilde{\mathcal{X}}, \tilde{\mathcal{X}}']) \tag{4.66a}$$

$$[W(\xi), W(\tilde{\mathcal{X}})] = 0 \tag{4.66b}$$

$$[W(\xi), W(\xi')] = W([\xi, \xi']) + W(i_\xi i_{\xi'} \mathbb{F})\,. \tag{4.66c}$$

*) The procedure of associating a Lie algebra to a given differential algebra is to be described in more detail in section IV.

Thus, by the change of generators $\mathfrak{X} \to \tilde{\mathfrak{X}}$, we have passed from a differential algebra which is associated to an ordinary Lie algebra to one which corresponds to a field dependent Lie algebra. In the latter the translations and susy transformations do not act on the supergauge transformations.

4. As mentioned before, the Russian formula for a gauge theory takes a slightly different form in the presence of space transformations[91),169)]. In fact the BRS equations for $\mathbb{A}$ and $\tilde{\mathfrak{X}}$ (eqs 4.58c,e) are equivalent to the formula

$$\begin{aligned} \mathcal{F} &= e^{-i_\xi} \mathbb{F} \\ &\equiv \mathbb{F} - i_\xi \mathbb{F} + \frac{1}{2} i_\xi i_\xi \mathbb{F} \end{aligned} \tag{4.67}$$

where

$$\mathcal{F} \equiv (d+s)(\mathbb{A} + \tilde{\mathfrak{X}}) + (\mathbb{A} + \tilde{\mathfrak{X}})(\mathbb{A} + \tilde{\mathfrak{X}}) .$$

With these notations the generalized transgression formula (4.35b) reads

$$\begin{aligned} J_3(\mathcal{F}, \mathcal{F}, \mathcal{F}) &= (d+s)\left[3 \int_0^1 dt\, J_3(\mathbb{A} + \tilde{\mathfrak{X}}, \mathcal{F}_t, \mathcal{F}_t)\right] \\ &\equiv (d+s)\, Q_5(\mathbb{A} + \tilde{\mathfrak{X}}) . \end{aligned} \tag{4.68}$$

Expanding the l.h.s. up to second order in ξ (and $\tilde{\mathfrak{X}}$) by insertion of (4.67),

$$J_3(\mathcal{F}, \mathcal{F}, \mathcal{F}) \simeq J_3(\mathbb{F}, \mathbb{F}, \mathbb{F}) - i_\xi J_3(\mathbb{F}, \mathbb{F}, \mathbb{F}) + \frac{1}{2} i_\xi i_\xi J_3(\mathbb{F}, \mathbb{F}, \mathbb{F})$$

and similarly the r.h.s.,

$$(d+s)\, Q_5(\mathbb{A} + \tilde{\mathfrak{X}}) \simeq dQ_5^0 + (s\, Q_5^0 + dQ_4^1) + (sQ_4^1 + dQ_3^2)$$

we obtain the modified descent equations

$$\begin{aligned} J_3(\mathbb{F}, \mathbb{F}, \mathbb{F}) &= dQ_5^0(\mathbb{A}) \\ sQ_5^0 + d\, Q_4^1 &= -i_\xi J_3(\mathbb{F}, \mathbb{F}, \mathbb{F}) \\ sQ_4^1 + d\, Q_3^2 &= \frac{1}{2} i_\xi i_\xi J_3(\mathbb{F}, \mathbb{F}, \mathbb{F}) . \end{aligned} \tag{4.69}$$

Contrary to the ordinary space-time situation[91),169)] the 6-form $J_3(\mathbb{F},\mathbb{F},\mathbb{F})$ does not vanish identically on (4+4)-dimensional superspace and therefore we do not obtain a solution of the WZ-consistency equations. However the constraints on the connection $\mathbb{A}$ can be used to throw the l.h.s. of equation (4.68) under the (d+s)-symbol by which method one gets an explicit expression for the susy chiral anomaly[26),136)].

5. By systematically applying superfield identities we have obtained simple formulae in this section without explicitly evaluating total derivatives of ghost fields. Failure to do so leads to sign ambiguities of the type encountered in the purely bosonic theory, see eq. (4.9). To elucidate their origin we consider the differentiation of an ordinary 1-superform:

$$dE^A = d(dz^M E_M{}^A) = dz^M dz^N \partial_N E_M{}^A$$

In this equation the quantity $dz^N\partial_N$ can be formally interpreted as a 1-form and therefore permutation with the monomial dz^M leads to the relation

$$dE^A = -(dz^N\partial_N)(dz^M E_M{}^A) = -(dz^N\partial_N) E^A$$
$$= -(E^B D_B) E^A .$$

Thus the correct differentiation rule of a superfield ξ^A with ghost number one (i.e. a quantity with total degree one) reads:

$$d\xi^A = -E^B(D_B\xi^A) \quad . \tag{4.70}$$

As emphasized above this kind of formulae has been avoided in our treatment.

(iii) Projection to space-time

Let us now consider the projection of the fields and equations from superspace to the usual space-time. For this purpose we specify to d=4 and N=1,2 for which cases there are well-known superspace formulations of the off-shell theory*)[154),65)]. Furthermore we choose the gauge vector representation for which $\mathbb{A}^\dagger = -\mathbb{A}$ and $\tilde{\mathfrak{X}}^\dagger = -\tilde{\mathfrak{X}}$ so that the superfields $\mathbb{A}_a$ and $\tilde{\mathfrak{X}}$ directly project to real space-time fields:

$$\mathbb{A}_a| \equiv -\frac{i}{2}v_a \quad , \quad \tilde{\mathfrak{X}}| \equiv \frac{i}{2}\tilde{v} \quad . \tag{4.71}$$

The classical action (3.8) for the real basis superfields Φ and $W^\alpha = \frac{i}{4}\mathbb{F}_{a\dot{\alpha}}(\bar{\sigma}^a)^{\dot{\alpha}\alpha}$ is invariant under the s-variations (4.58). Indeed, by taking the

*) The final result for the N=2 theory will be given in part V.

hermitean conjugate of

$$s\Phi = \tilde{\mathcal{X}}\Phi - \xi^{B}\mathcal{D}_{B}\Phi$$

and by using the fact that $(\xi_b)^* = \xi_b$ and $(\xi_\beta)^* = \bar{\xi}_{\dot{\beta}}$ for the susy generating vector field (1.5) we obtain

$$s\Phi^{\dagger} = -\Phi^{\dagger}\tilde{\mathcal{X}} - (\mathcal{D}^{B}\Phi)^{\dagger}\xi_{B} .$$

Hence

$$\begin{aligned} s\,S_{\Phi} &= \int d^8z\; s(\Phi^{\dagger}\Phi) = \int d^8z\left[\Phi^{\dagger}(s\Phi) + (s\Phi^{\dagger})\Phi\right] \\ &= -\int d^8z\left[\Phi^{\dagger}\xi^{B}(\mathcal{D}_{B}\Phi) + (\mathcal{D}^{B}\Phi^{\dagger})\xi_{B}\Phi\right] \\ &= -\int d^8z\;\mathcal{D}^{B}(\Phi^{\dagger}\xi_{B}\Phi) = -\int d^8z\; D^{B}(\Phi^{\dagger}\xi_{B}\Phi) \quad = 0 \end{aligned}$$

where we used the relation $D^{B}\xi_{B} = 0$ and the supergauge invariance of $\Phi^{\dagger}\xi_{B}\Phi$. The s-invariance of the gauge field contribution follows in a similar way by using the cyclicity property of the supertrace.

In order to obtain the transformation laws of the component fields <u>we define</u>

$$s(Q|) := (sQ)| \tag{4.72}$$

<u>for all superfields Q</u>. In the x-space equations resulting from this projection procedure the translation contributions appear in a gauge covariant form (e.g. (4.58a) projects to

$$s(\Phi|) = \frac{i}{2}\tilde{v}\,\Phi| - a^{m}\mathcal{D}_{m}(\Phi|) - \zeta^{\alpha}(\mathcal{D}_{\alpha}\Phi)| \qquad),$$

but the familiar form of the component field algebra can easily be recovered by the <u>change of variables</u> $v := \tilde{v} - a^{m}v_{m}$.

For later reference we collect the so-derived relations*)

$$s(\Phi|) = \frac{i}{2}v\,\Phi| - (a\cdot\partial)\Phi| - \zeta^{\alpha}(\mathcal{D}_{\alpha}\Phi)| \tag{4.73a}$$

$$s\,v_{a} = \mathcal{D}_{a}v - (a\cdot\partial)v_{a} - 2i\,\zeta^{\beta}\,\mathbb{F}_{\beta a}| \tag{4.73b}$$

$$s(\mathbb{F}_{BA}|) = -\frac{i}{2}\left[\mathbb{F}_{BA}|, v\right] - (a\cdot\partial)\mathbb{F}_{BA}| - \zeta^{\gamma}(\mathcal{D}_{\gamma}\mathbb{F}_{BA})| \tag{4.73c}$$

---

*) Note that $s\xi^{M} = -\xi^{N}(\partial_{N}\xi^{M})$.

$$s v = -\frac{i}{2} v v - (a\cdot\partial) v - i \xi^{\alpha} \xi^{\beta} \mathbb{F}_{\underline{\beta}\underline{\alpha}}| - 2i(\xi\sigma^m\bar{\xi}) v_m \quad (4.73d)$$

$$s a^m = 2i\,(\xi\sigma^m\bar{\xi}) \quad (4.73e)$$

$$s \xi^{\mu} = 0 = s \bar{\xi}_{\dot{\mu}} \quad (4.73f)$$

The first equation holds in particular for $\mathcal{D}_{\alpha}\Phi$ and $\mathcal{D}^{\alpha}\mathcal{D}_{\alpha}\Phi$ while the third also applies to the superfields $W^{\alpha} \equiv \frac{i}{4}\mathbb{F}_{a\dot{\alpha}}(\bar{\sigma}^a)^{\dot{\alpha}\alpha}$ and $\mathcal{D}^{\alpha}W_{\alpha}$.

Now insertion of the components (3.1),(3.3) of the superfields W_{α}, Φ and implementation of the N=1 constraints $\mathbb{F}_{\underline{\alpha}\underline{\beta}} = 0$ directly leads to the BRS component field algebra in the WZ-gauge:

$$s A = \frac{i}{2} v A - (a\cdot\partial) A - \sqrt{2}\,\xi\psi \quad (4.74a)$$

$$s \psi = \frac{i}{2} v \psi - (a\cdot\partial)\psi - i\sqrt{2}(\sigma^m\bar{\xi})\mathcal{D}_m A - \sqrt{2}\,\xi F \quad (4.74b)$$

$$s F = \frac{i}{2} v F - (a\cdot\partial) F - i\sqrt{2}\,\bar{\xi}\bar{\sigma}^m \mathcal{D}_m\psi - i(\bar{\xi}\bar{\lambda})A \quad (4.74c)$$

$$s v_m = \mathcal{D}_m v - (a\cdot\partial) v_m - i[\xi\sigma_m\bar{\lambda} + \bar{\xi}\bar{\sigma}_m\lambda] \quad (4.74d)$$

$$s\lambda = \frac{i}{2}[v,\lambda] - (a\cdot\partial)\lambda - (\sigma^{mn}\xi) v_{mn} - i\xi D \quad (4.74e)$$

$$s D = \frac{i}{2}[v, D] - (a\cdot\partial) D + \xi\sigma^m(\mathcal{D}_m\bar{\lambda}) - \bar{\xi}\bar{\sigma}^m(\mathcal{D}\lambda) \quad (4.74f)$$

$$s v = -\frac{i}{2} v v - (a\cdot\partial) v - 2i(\xi\sigma^m\bar{\xi}) v_m \quad (4.74g)$$

$$s a^m = 2i\,(\xi\sigma^m\bar{\xi}) \quad (4.74h)$$

$$s\xi = 0 = s\bar{\xi} \quad . \quad (4.74i)$$

The nilpotency of these transformations follows automatically from the definition (4.72) and was explicitly verified in order to check the consistency of the whole approach. Concerning this point we remark the following. At the superfield level the total differential d acts from the right and is not sensitive to the statistics degree of the superfields and forms ; on the other hand the partial differential operators ∂_M act from the left as derivations and are affected by the statistics degree of the superfields:

$$\partial_M(\xi^N \eta^L) = (\partial_M\xi^N)\eta^L + (-)^{mn}\xi^N(\partial_M\eta^L) \quad .$$

In the absence of ghost fields these superspace conventions lead to the usual product rules for the partial derivatives of space-time fields. Though in the present situation these conventions induce non-standard differentiation rules for partial derivatives when acting on products of x-space fields involving ghost parameters $(v, a^m$ or $\zeta)$. Also, as a consequence of the bigrading (4.53), the graded commutator $[\mathbb{F},\mathfrak{X}]$ projects to a commutator $[\lambda_\alpha, v]$ instead of an anticommutator for the two Grassmann variables λ_α, v *). Thus, although our geometric approach is consistent in itself, it is advisable to modify appropriately the superspace induced conventions in any application of the final component field algebra (just add the ghost number to the statistics degree and suppose that the BRS operator acts from the left as in the bosonic case).

The transformations (4.74) can readily be rewritten in terms of Majorana 4-spinors, though such a notation only makes sense for the matter fields, if these are in a real representation of the gauge group.

(iv) Summary

Our approach can be schematically summarized by the following diagram:

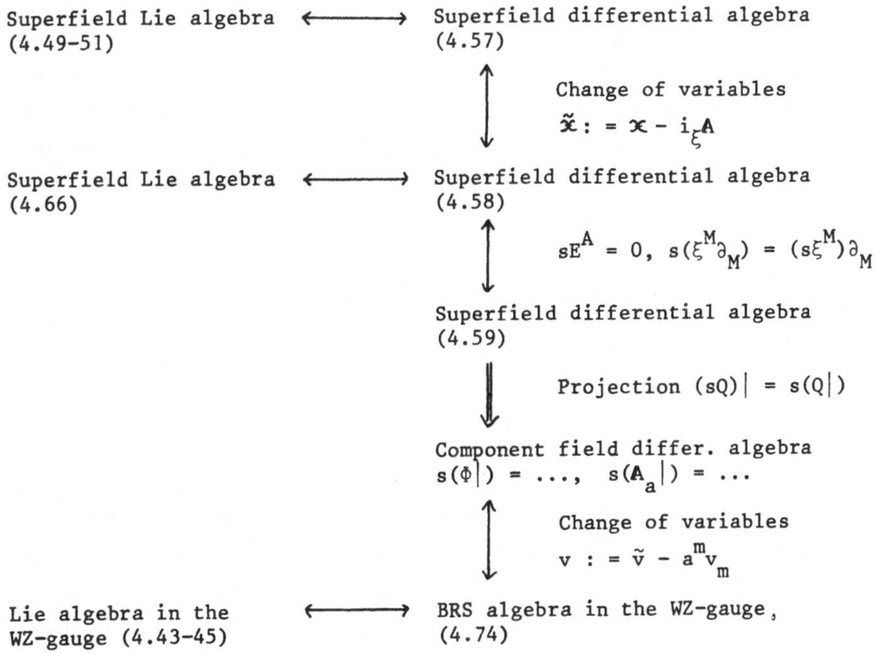

Superfield Lie algebra (4.49-51) ⟷ Superfield differential algebra (4.57)

↕ Change of variables $\tilde{\mathfrak{X}} := \mathfrak{X} - i_\xi \mathbb{A}$

Superfield Lie algebra (4.66) ⟷ Superfield differential algebra (4.58)

↕ $sE^A = 0, \; s(\xi^M \partial_M) = (s\xi^M)\partial_M$

Superfield differential algebra (4.59)

⇓ Projection $(sQ)| = s(Q|)$

Component field differ. algebra $s(\Phi|) = \ldots, \; s(\mathbb{A}_a|) = \ldots$

↕ Change of variables $v := \tilde{v} - a^m v_m$

Lie algebra in the WZ-gauge (4.43-45) ⟷ BRS algebra in the WZ-gauge, (4.74)

---

*) This consequence of the geometric grading (4.53) has been pointed out before by J. Thierry-Mieg and Y. Ne'eman[142]. These authors also discussed the implications of the $Z(2)\times Z(2)$-grading for the quantum theory.

If we suppose that the constraints are implemented while projecting to space-time, all steps in the diagram are reversible except the projection process.

(v) Structure of the differential algebra and of the associated Lie algebra*)

On the classical fields v_m, λ_α and D the s-operation (4.74) corresponds to the sum of the infinitesimal variations (4.43) and therefore the Lie algebra associated to the differential algebra (4.74) is given by the equations (4.43-45). The question to be addressed in this section is whether or not we can pass from the field dependent Lie algebra (4.43-45) or equivalently from the differential algebra (4.74) to other algebras which do not depend on the dynamical fields of the theory. (We recall that the change of variables $\tilde{\mathfrak{X}} := \mathfrak{X} - i_\xi A$ considered before the projection process does exactly this job for the different superfield algebras.)

For the differential algebra (4.74) our question can be formulated more precisely as follows : Is it possible to give a standard (field independent) form to the s-transformation of the ghost field v,

$$sv = -\frac{i}{2} vv - (a \cdot \partial) v - 2i \left(\zeta \sigma^m \bar{\zeta} \right) v_m \tag{4.75}$$

by an appropriate (field dependent) change of variables?

To answer this question we take into account the gradings of the fields as well as their mass dimension, i.e. an additional degree which is not affected by the BRS operator. The only possibility for eliminating the v_m-dependent term from equation (4.75) is by a change of the generator v. The latter is a real Lie algebra valued scalar field with total degree one and mass dimension $[v] = 0$. The only other parameters of degree one occuring in the differential algebra are a^m and ζ while the mass dimensions of all involved fields and parameters are given by

$$[a^m] = -1 \quad , \quad [\zeta] = -\frac{1}{2} \quad , \quad [v] = 0$$

$$[v_m] = 1 \quad , \quad [\lambda_\alpha] = 3/2 \quad , \quad [D] = 2 = [v_{mn}] .$$

From these facts it immediately follows that the only change of variables which is compatible with the above restrictions is $\tilde{v} = v + k\, a^m v_m$ (with k a real number). But this is precisely the type of transformation considered above to cast the translation contributions into a familiar form: it introduces terms depending on λ and $\bar{\lambda}$, e.g. for $k = 1$

$$s\tilde{v} = -\frac{i}{2} \tilde{v}\tilde{v} - \frac{1}{2} a^m a^n v_{mn} - i a^m \left[\zeta \sigma_m \bar{\lambda} + \bar{\zeta} \bar{\sigma}_m \lambda \right] .$$

---

*) Without a loss of generality we restrict the following considerations to the gauge multiplet.

Thus, <u>without introducing additional fields, the BRS algebra in the WZ-gauge cannot be given a field independent form</u>. A posteriori the "anomalies in the WZ-gauge" correspond to cohomology classes of a field dependent Lie algebra.

The origin of this negative result is easy to spot. In the process of projection to space-time the supergeometry has been destroyed by the elimination of the higher components of the superfield $\tilde{\mathfrak{X}}$, i.e. the supersymmetric partners of $\frac{i}{2}\tilde{v} = \tilde{\mathfrak{X}}|$. While the translations $W(a)$ still act on the ordinary gauge transformations $W(v)$ in the component field algebra (4.45), an action of the susy transformations on the real gauge algebra $\{W(v)\}$ cannot be defined without introducing additional fields. Thus we do not have anymore the semi-direct sum structure (4.52) which characterized the superfield Lie algebra (4.49-51).

(v) <u>Conclusion</u>

The strong reality condition $\mathbb{A}^{\dagger} = -\mathbb{A}$ characterizing the gauge vector representation allows a geometric formulation of classical SYM-theories which projects without any supergauge fixing to a simple space-time theory. But, analogously to the WZ-gauge fixing in the chiral basis, this process destroys the supergeometry and in particular the mathematical structure of the BRS differential algebra ; the resulting component field algebras are necessarily field dependent.

V. GEOMETRY OF EXTENDED SUPERSYMMETRY

The purpose of these final chapters is to generalize the previously presented formalism and results to the case of extended supersymmetry[119),*)]. As far as the geometry of rigid superspace is concerned, this can be done in a straightforward way ; essentially it amounts to the addition of a few supplementary indices representing the internal symmetry group U(N). On the other hand the constraints on the YM-superconnection become considerably more complicated and so does their solution. In fact it is only for the solution of the constrained Bianchi identities of the N=2 theory that one recovers a N=1 like situation. But even in this "simple" case there is no known analytic expression for the explicit solution of the constraints and therefore the quantization of these theories represents a major problem. It turns out that the appropriate arena for the formulation and quantization of extended SYM-theories is not given by the U(N) rigid superspace, but the so-called harmonic superspace which was introduced a few years ago by A. Galperin et al.[55)] and which allows to solve the constraints explicitly in complete analogy to the N=1 case.

V.I Geometry of rigid U(N)-superspace

The 4-dimensional N-extended susy algebra (without a central extension) is characterized by the following commutation relations:

$$\begin{aligned} &\{ Q^i_\alpha , \bar{Q}_{\dot\beta j} \} = 2 \left(\delta^i_j \, \sigma^m_{\alpha\dot\beta} \right) P_m \\ &\{ Q^i_\alpha , Q^j_\beta \} = \{ \bar{Q}_{\dot\alpha i} , \bar{Q}_{\dot\beta j} \} = 0 \\ &[P_m , Q^i_\alpha] = [P_m , \bar{Q}_{\dot\alpha i}] = [P_m , P_n] = 0 \quad . \end{aligned} \tag{5.1}$$

Here the indices $i,j \in \{1,\ldots,N\}$ represent the internal symmetry group $U(N)$[**)]. This algebra can be formally exponentiated by using real numbers x^m and global Weyl spinors θ^α_i, $\bar\theta^j_{\dot\alpha}$ which transform under the fundamental representation of U(N) and its complex conjugate, respectively:

---

*) A general introduction to this subject can be found in the reviews of M.Sohnius[128)] and P. West[155),156)] (see also ref. 127)).

**) These indices can be lowered (resp. raised) by the totally antisymmetric tensor $\varepsilon_{\underbrace{ij\ldots}_{N}}$ (resp. its inverse). Complex conjugation is discussed in Appendix 3.

$$G(x^m, \theta_i^\alpha, \bar{\theta}_{\dot{\alpha}}^j) = \exp[-ixP + i\,\theta_k Q^k + i\,\bar{\theta}^k \bar{Q}_k] \tag{5.2}$$

From the structure of the algebra (5.1) it is clear that the geometric framework of simple and extended supersymmetry is quite similar : the spinorial indices $(\alpha,\dot{\alpha})$ of the N=1 theory are to be replaced by double indices $(\alpha i,\dot{\alpha} j)$ and the Pauli matrices $\sigma^m_{\alpha\dot{\beta}}$ occuring in the structure constants of the susy algebra become $\delta^i{}_j \sigma^m_{\alpha\dot{\beta}}$. By making the appropriate substitutions one can obtain all the canonical quantities characterizing rigid U(N)-superspace (which is defined in analogy to the N=1 space). For these reasons we limit ourselves to a collection of the important relations and we refer to part I for the general theory.

Natural parametrization of U(N)-superspace:

$$(z^M) = (x^m, \theta_i^\mu, \bar{\theta}_{\dot{\mu}}^j), \qquad (\theta_i^\mu)^* = \bar{\theta}^{\dot{\mu} i} \tag{5.3}$$

Natural tangent space basis:

$$(\partial_M) = (\partial_m, \partial_\mu^i, \bar{\partial}_j^{\dot{\mu}}) = \left(\frac{\partial}{\partial x^m}, \frac{\partial}{\partial \theta_i^\mu}, \frac{\partial}{\partial \bar{\theta}_{\dot{\mu}}^j}\right) \tag{5.4}$$

Supersymmetry generators:

$$\begin{aligned} Q_\alpha^k &= \frac{\partial}{\partial \theta_k^\alpha} - i(\sigma^m \bar{\theta}^k)_\alpha \partial_m \\ \bar{Q}_k^{\dot{\alpha}} &= \frac{\partial}{\partial \bar{\theta}_{\dot{\alpha}}^k} - i\,\theta_k^\alpha \sigma^m_{\alpha\dot{\beta}} \varepsilon^{\dot{\beta}\dot{\alpha}} \partial_m \end{aligned} \tag{5.5}$$

Left-invariant derivatives:

$$\begin{aligned} D_\alpha^k &= \frac{\partial}{\partial \theta_k^\alpha} + i(\sigma^m \bar{\theta}^k)_\alpha \partial_m \\ \bar{D}_{\dot{\alpha} k} &= -\frac{\partial}{\partial \bar{\theta}^{\dot{\alpha} k}} - i\,\theta_k^\alpha \sigma^m_{\alpha\dot{\alpha}} \partial_m \end{aligned} \tag{5.6}$$

Left-invariant vielbein forms $E^A = dz^M E_M{}^A$:

$$\begin{aligned} E^a &= dx^a - i(d\theta_k)\sigma^a \bar{\theta}^k + i\,\theta_k \sigma^a d\bar{\theta}^k \\ E_k^\alpha &= d\theta_k^\alpha \\ E_{\dot{\alpha}}^k &= d\bar{\theta}_{\dot{\alpha}}^k \end{aligned} \tag{5.7}$$

The corresponding vielbein fields:

$$\left(E_M{}^A\right)=\begin{bmatrix} E_m{}^a=\delta_m{}^\alpha & E_m{}^\alpha{}_j=0 & E_{m\dot\alpha}{}^j=0 \\ E_\mu^{k\,a}=-i\,\sigma^a_{\mu\dot\mu}\,\bar\theta^{\dot\mu k} & E^{k\alpha}_{\mu\;j}=\delta_\mu{}^\alpha\,\delta_j{}^k & E^{kj}_{\mu\dot\alpha}=0 \\ E^{\dot\mu a}_k=-i\,\theta^\rho_k\,\sigma^a_{\rho\dot\nu}\,\varepsilon^{\dot\nu\dot\mu} & E^{\dot\mu\alpha}_{k\,j}=0 & E^{\dot\mu j}_{k\dot\alpha}=\delta_{\dot\alpha}{}^{\dot\mu}\,\delta_k{}^j \end{bmatrix} \tag{5.8a}$$

resp.

$$\left(E_A{}^M\right)=\begin{bmatrix} E_a{}^m=\delta_a{}^m & E_a{}^\mu{}_j=0 & E_{a\dot\mu}{}^k=0 \\ E^{k\,m}_\alpha=i\,\sigma^m_{\alpha\dot\beta}\,\bar\theta^{\dot\beta k} & E^{k\mu}_{\alpha\,j}=\delta_\alpha{}^\mu\,\delta_j{}^k & E^{kj}_{\alpha\dot\mu}=0 \\ E^{\dot\alpha\,m}_k=i\,\theta^\alpha_k\,\sigma^m_{\alpha\dot\beta}\,\varepsilon^{\dot\beta\dot\alpha} & E^{\dot\alpha\mu}_{k\,j}=0 & E^{\dot\alpha j}_{k\dot\mu}=\delta_{\dot\mu}{}^{\dot\alpha}\,\delta_k{}^j \end{bmatrix} \tag{5.8b}$$

Torsion $T^A = dE^A$ of the canonical linear connection $(\phi \equiv 0)$:

$$T^{\;k}_{\alpha\;\;\dot\beta j}{}^{c} = \left(2i\,\sigma^c_{\alpha\dot\beta}\right)\delta_j{}^k = T_{\dot\beta j}{}^{k}{}_{\alpha}{}^{c} \tag{5.9}$$

Susy transformations (and ordinary translations):

$$\begin{aligned} x'^m &= x^m + a^m + i\left(\theta_k\sigma^m\bar\zeta^k\right) - i\left(\zeta_k\sigma^m\bar\theta^k\right) \\ \theta'_k &= \theta_k + \zeta_k \\ \bar\theta'^k &= \bar\theta^k + \bar\zeta^k \end{aligned} \tag{5.10}$$

Susy generating vector field (satisfying $L_\xi E^A = 0$):

$$\left(\xi^M\right) = \left(a^m + i\left(\theta_k\sigma^m\bar\zeta^k\right) - i\left(\zeta_k\sigma^m\bar\theta^k\right),\ \zeta_k\ ,\ \bar\zeta^k\right) \tag{5.11}$$

In the N=2 case one can construct a SU(2)-covariant 4-component spinor λ^i from the 2-component spinors (λ^i_α) and $(\bar\lambda^{\dot\alpha i}) \equiv (\lambda^\alpha_i)^+$: the <u>symplectic Majorana spinor</u> λ^i is defined by

$$\lambda^i \equiv \begin{pmatrix} -i\,\lambda^i_\alpha \\ \bar\lambda^{\dot\alpha i} \end{pmatrix} \qquad \Longrightarrow \qquad \bar\lambda_i = \left(\lambda^\alpha_i\ ,\ -i\,\bar\lambda_{\dot\alpha i}\right)$$

and satisfies the symplectic reality condition[119),128)]

$$\lambda^i = - \varepsilon^{ij} (\gamma^5 C) (\bar{\lambda}_j)^t$$

Here

$$\gamma^5 = \begin{bmatrix} -i \mathbb{1}_2 & \\ & i \mathbb{1}_2 \end{bmatrix} \quad \text{and} \quad C = \begin{bmatrix} -\varepsilon_{\alpha\beta} & \\ & -\varepsilon^{\dot\alpha\dot\beta} \end{bmatrix} .$$

V.2 General structure of extended SYM-theories

(i) Geometric framework

The geometric set-up for extended SYM-theories is the same as for the N=1 theory. Thus one starts with a superconnection

$$\mathbb{A} = E^A \mathbb{A}_A = E^a \mathbb{A}_a + E^\alpha_k \mathbb{A}^k_\alpha + E^k_{\dot\alpha} \mathbb{A}^{\dot\alpha}_k \tag{5.12}$$

and a supercurvature

$$\mathbb{F} = d\mathbb{A} + \mathbb{A}\mathbb{A} = \frac{1}{2} E^A E^B \mathbb{F}_{BA} \tag{5.13}$$

with

$$\mathbb{F}_{BA} = D_B \mathbb{A}_A - (-)^{ab} D_A \mathbb{A}_B - [\mathbb{A}_A , \mathbb{A}_B] + T_{AB}{}^C \mathbb{A}_C , \tag{5.14}$$

for instance:

$$\mathbb{F}^{ij}_{\alpha\beta} = D^i_\alpha \mathbb{A}^j_\beta + D^j_\beta \mathbb{A}^i_\alpha - \{ \mathbb{A}^i_\alpha , \mathbb{A}^j_\beta \}$$

$$\mathbb{F}_{\dot\alpha i \, \dot\beta j} = \bar{D}_{\dot\alpha i} \mathbb{A}_{\dot\beta j} + \bar{D}_{\dot\beta j} \mathbb{A}_{\dot\alpha i} - \{ \mathbb{A}_{\dot\alpha i} , \mathbb{A}_{\dot\beta j} \}$$

$$\mathbb{F}^k_{\alpha}{}_{\dot\beta j} = D^k_\alpha \mathbb{A}_{\dot\beta j} + \bar{D}_{\dot\beta j} \mathbb{A}^k_\alpha - \{ \mathbb{A}^k_\alpha , \mathbb{A}_{\dot\beta j} \} + (2i \, \sigma^c_{\alpha\dot\beta} \, \delta_j{}^k) \, \mathbb{A}_c .$$

We mention the following useful identity satisfied by the gauge covariant derivatives $\mathcal{D}^i_\alpha$:

$$\mathcal{D}^i_\alpha \mathcal{D}^j_\beta = \frac{1}{2} \left(\varepsilon_{\alpha\beta} \mathcal{D}^{ij} + \varepsilon^{ij} \mathcal{D}_{\alpha\beta} - \mathbb{F}^{ij}_{\alpha\beta} \right) \tag{5.15}$$

where

$$\mathcal{D}^{ij} = \mathcal{D}^{\alpha i}\mathcal{D}^{j}_{\alpha} \qquad \text{and} \qquad \mathcal{D}_{\alpha\beta} = \mathcal{D}^{i}_{\alpha}\mathcal{D}_{\beta i} \quad .$$

(similarly for $\bar{\mathcal{D}}_{\dot\alpha i}$).

(ii) Constraints

The most difficult task in the superspace formulation of SYM-theories is the determination of susy and supergauge covariant constraint equations which do not lead to equations of motion or to a pure gauge theory. For N=2 SYM such constraints have been proposed by R. Grimm, M. Sohnius and J. Wess[65]:

$$\mathbb{F}^{i\;\;j}_{(\alpha\;\beta)} \equiv \mathbb{F}^{i\;j}_{\alpha\beta} + \mathbb{F}^{i\;j}_{\beta\alpha} = 0 \qquad \left(\Longleftrightarrow \;\mathbb{F}^{(i\;j)}_{\alpha\;\beta} = 0\right)$$

$$\mathbb{F}_{(\dot\alpha i\;\dot\beta)j} \equiv \mathbb{F}_{\dot\alpha i\dot\beta j} + \mathbb{F}_{\dot\beta i\dot\alpha j} = 0 \qquad \left(\Longleftrightarrow \;\mathbb{F}_{\dot\alpha(i\;\dot\beta j)} = 0\right)$$

$$\mathbb{F}^{i}_{\alpha\;\dot\beta j} = 0 \qquad (5.16)$$

Reality condition : $\mathbb{F}^{\dagger} \cong \mathbb{F}$ (see (2.29)).

Later[58] it was realized that these equations correspond to a redefinition constraint ($\mathbb{F}^{i}_{\alpha\dot\beta j} = 0$) and a set of representation - preserving constraints for the scalar hypermultiplet[49] which is the smallest multiplet in N=2 supersymmetry (with a central charge).

For $N \geq 3$ the conditions (5.16) are too stringent and put the theory on-shell[126]. Indeed for N=3 and N=4 SYM in four dimensions the constraint equations (5.16) are completely equivalent to the supersymmetric field equations, as was first recognized by E. Witten[158] and later explicitly proven for N=3 by J. Harnad et al.[71] (See also the references 2) and 72) for the equivalence between the constraints and the field equations in d=10, N=1 SYM which leads to the d=4, N=4 theory by dimensional reduction[30),177)].)

(iii) Solution of the constraints and Bianchi identities

For N=2 the Bianchi identities subject to the constraints (5.16) have been solved in ref. 65)*). The general solution is given by two Lie algebra valued superfields W and $\bar{W}$ satisfying

---

*) We use the formulation given in ref. 126).

$$0 = \mathcal{D}^{i}_{\alpha} \bar{W} = \bar{\mathcal{D}}_{\dot{\alpha} i} W = \mathcal{D} \varepsilon \vec{\tau} g \mathcal{D} W - \bar{\mathcal{D}} \varepsilon g \vec{\tau} \bar{\mathcal{D}} \bar{W}$$

$$W^{\dagger} \cong \bar{W} . \tag{5.17a}$$

Here g^{ij} stands for the ε-symbol in the SU(2)-space and $\vec{\tau}_{i}{}^{j}$ represents the Pauli matrices. For the components of the supercurvature one has the expressions

$$\begin{aligned}
\mathbb{F}^{i\,j}_{\alpha\,\beta} &= \varepsilon_{\alpha\beta}\, g^{ij}\, \bar{W} \\
\mathbb{F}_{\dot{\alpha} i\, \dot{\beta} j} &= \varepsilon_{\dot{\alpha}\dot{\beta}}\, g_{ij}\, W \\
\mathbb{F}_{a\,\alpha}{}^{i} &= -\frac{i}{4} (\sigma_a)_{\alpha\dot{\beta}} (\bar{\mathcal{D}}^{\dot{\beta} i} \bar{W}) \\
\mathbb{F}_{a\,\dot{\alpha} i} &= \frac{i}{4}\, \varepsilon_{\dot{\alpha}\dot{\beta}}\, \bar{\sigma}_a{}^{\dot{\beta}\alpha} (\mathcal{D}_{\alpha i} W) \\
\mathbb{F}_{ab} &= -\frac{1}{16} \left(\bar{\mathcal{D}}_{\dot{\alpha} i} (\bar{\sigma}_{ab})^{\dot{\alpha}}{}_{\dot{\beta}}\, \bar{\mathcal{D}}^{\dot{\beta} i} \bar{W} + \mathcal{D}^{\alpha}_{i} (\sigma_{ab})_{\alpha}{}^{\beta} (\mathcal{D}^{i}_{\beta} W) \right) .
\end{aligned} \tag{5.17b}$$

In the chiral/anti-chiral representation of the N=1 theory, the constraints $\mathbb{F}_{\underline{\alpha\beta}} = 0$, $\mathbb{A}^{\dagger} \cong - \mathbb{A}$ could be solved explicitly in terms of a single superfield V. In the N=2 case a solution of the constraints (5.16) is rather complicated and an analytic solution is only known for the linearized theory[98] (see refs 87), 156) and references therein). It is this fact which renders the superspace quantization of N=2 SYM quite complicated[75]. These problems do not arise, if the theory is formulated in N=2 harmonic superspace[55]: here the constraints (5.16) can be solved explicitly in analogy to the N=1 case and the different types of solutions lead to a similar triangular structure as the one encountered in part II (ref. 55), see also ref. 38)).

The solution for $N \geq 3$ has been discussed in ref. 126) and will not be considered here, since it only leads to the equations of motion ; for an unconstrained off-shell formulation of the N=3 theory in harmonic superspace we refer to the pioneering work of the Russian group[55].

(iv) The gauge real representation (N=2)

In the gauge real representation we have $W^{\dagger} = \bar{W}$ and the field content of the theory can be specified by

$$\mathbb{A}_a | = -\frac{i}{2} v_a \tag{5.18a}$$

and the following independent gauge covariant projections of W :

$$W| \equiv \frac{1}{2} A \quad , \quad \mathcal{D}^i_\alpha W| \equiv -\frac{i}{2} \lambda^i_\alpha \quad , \quad \mathcal{D}\varepsilon \vec{\tau} g \mathcal{D} W| \equiv \frac{1}{2} \vec{D} \; . \tag{5.18b}$$

As pointed out in section III.2 (i) the "susy transformations" of these component fields directly follow from the relations (3.4),(3.5)*):

$$\begin{aligned}
\delta_\zeta A &= - i \zeta_k \lambda^k \\
\delta_\zeta v_m &= -\frac{i}{4} \left(\zeta_k \sigma_m \bar{\lambda}^k + \bar{\zeta}^k \bar{\sigma}_m \lambda_k \right) \\
\delta_\zeta \lambda^k &= -4 (\sigma^{mn} \zeta^k) v_{mn} + i \zeta^j (\vec{\tau}_j{}^k \cdot \vec{D}) - 2 (\sigma^m \bar{\zeta}^k) \mathcal{D}_m A + \frac{i}{4} \zeta^k [\bar{A}, A] \\
\delta_\zeta \vec{D} &= -4 \left[\zeta^j \vec{\tau}_j{}^k \sigma^m (\mathcal{D}_m \bar{\lambda}_k) + \bar{\zeta}^j \vec{\tau}_j{}^k \bar{\sigma}^m (\mathcal{D}_m \lambda_k) \right] \\
&\quad - i \left[\zeta^j \vec{\tau}_j{}^k \lambda_k , \bar{A} \right] + i \left[\bar{\zeta}^j \vec{\tau}_j{}^k \bar{\lambda}_k , A \right] \; .
\end{aligned} \tag{5.19}$$

As for the N=1 theory we are in a WZ-gauge like situation and the commutator of two such susy transformations reads:

$$\begin{aligned}
\left[\zeta^\alpha \mathcal{D}_\alpha , \eta^\beta \mathcal{D}_\beta \right] \phi &= -2i \left(\zeta_k \sigma^m \bar{\eta}^k - \eta_k \sigma^m \bar{\zeta}^k \right) \left(\partial_m - A_m \right) \phi \\
&\quad + \left[(\zeta_k \eta^k) \bar{W} - (\bar{\zeta}^k \bar{\eta}_k) W \right] \phi \; .
\end{aligned}$$

By taking the lowest component of this equation we see that the commutator of two "susy transformations" corresponds to an ordinary translation with parameter $2i(\zeta_k \sigma^m \bar{\eta}^k - \eta_k \sigma^m \bar{\zeta}^k)$ and a gauge transformation with field dependent parameter

$$v = 2i \left(\zeta_k \sigma^m \bar{\eta}^k - \eta_k \sigma^m \bar{\zeta}^k \right) v_m + i \left[(\zeta_k \eta^k) \bar{A} - (\bar{\zeta}^k \bar{\eta}_k) A \right] \; . \tag{5.20}$$

We remark that the isotriplet $\vec{D}$ could also be described by the field[178)]

$$\left[\mathcal{D}^{\alpha i} , \mathcal{D}^j_\alpha \right] W| \equiv \frac{1}{2} C^{ij}$$

which corresponds to the symmetric part of the SU(2)-representation product $2 \otimes 2$.

---

*) Note that $\vec{\tau}_i{}^k \varepsilon_{kj} = \vec{\tau}_j{}^k \varepsilon_{ki}$.

V.3 N=2 BRS algebra in the WZ-gauge

Along the lines of section IV.2 (ii) we can immediately set up this algebra by considering the susy transformations (5.19) as well as the infinitesimal translations and gauge transformations given by

$$\begin{aligned}
\delta_a A &= -(a\cdot\partial)A & \delta_v A &= \tfrac{i}{2}[v, A] \\
\delta_a v_m &= -(a\cdot\partial)v_m & \delta_v v_m &= \mathcal{D}_m v \\
\delta_a \lambda^k &= -(a\cdot\partial)\lambda^k & \delta_v \lambda^k &= \tfrac{i}{2}[v, \lambda^k] \\
\delta_a \vec{D} &= -(a\cdot\partial)\vec{D} & \delta_v \vec{D} &= \tfrac{i}{2}[v, \vec{D}]
\end{aligned} \tag{5.21}$$

The field dependent parameter (5.20) yields a contribution

$$-2i(\xi_k\sigma^m\bar{\xi}^k)v_m - \tfrac{i}{2}\left[(\xi_k\xi^k)\bar{A} - (\bar{\xi}^k\bar{\xi}_k)A\right]$$

to sv so that we obtain the <u>differential algebra</u>:

$$sA = \tfrac{i}{2}[v, A] - (a\cdot\partial)A + i\xi_k\lambda^k \tag{5.22a}$$

$$sv_m = \mathcal{D}_m v - (a\cdot\partial)v_m + \tfrac{i}{4}(\xi_k\sigma_m\bar{\lambda}^k + \bar{\xi}^k\bar{\sigma}_m\lambda_k) \tag{5.22b}$$

$$\begin{aligned} s\lambda^k = \tfrac{i}{2}[v, \lambda^k] &- (a\cdot\partial)\lambda^k + 4(\sigma^{mn}\xi^k)v_{mn} - i\xi^j(\vec{\tau}_j{}^k\cdot\vec{D}) \\ &+ 2(\sigma^m\bar{\xi}^k)\mathcal{D}_m A - \tfrac{i}{4}\xi^k[\bar{A}, A] \end{aligned} \tag{5.22c}$$

$$\begin{aligned} s\vec{D} = \tfrac{i}{2}[v, \vec{D}] &- (a\cdot\partial)\vec{D} + 4\left[\xi^j\vec{\tau}_j{}^k\sigma^m(\mathcal{D}_m\bar{\lambda}_k) + \bar{\xi}^j\vec{\tau}_j{}^k\bar{\sigma}^m(\mathcal{D}_m\lambda_k)\right] \\ &+ i[\xi^j\vec{\tau}_j{}^k\lambda_k, \bar{A}] - i[\bar{\xi}^j\vec{\tau}_j{}^k\bar{\lambda}_k, A] \end{aligned} \tag{5.22d}$$

$$sv = -\tfrac{i}{2}vv - (a\cdot\partial)v - 2i(\xi_k\sigma^m\bar{\xi}^k)v_m - \tfrac{i}{2}\left[(\xi_k\xi^k)\bar{A} - (\bar{\xi}^k\bar{\xi}_k)A\right] \tag{5.22e}$$

$$sa^m = 2i(\xi_k\sigma^m\bar{\xi}^k) \tag{5.22f}$$

$$s\xi^k = 0 = s\bar{\xi}_k \quad . \tag{5.22g}$$

This is also the result following from the eqs (4.73b-d)(and (4.58f),(5.11)) upon implementation of the N=2 constraints (5.16). (Because of its complete covariance, eq.(4.73c) holds for all the component fields introduced in (5.18b).)

APPENDIX 1 : SUPERSPACE CONVENTIONS AND NOTATIONS (for N=1, d=4)[154)]

Minkowski metric:

$$(\eta_{mn}) = \mathrm{diag}(-1,1,1,1) \qquad m,n \in \{0,1,2,3\}$$

Sigma matrices:

$$\sigma^0 = \begin{bmatrix} -1 & 0 \\ 0 & -1 \end{bmatrix}, \quad \sigma^1 = \begin{bmatrix} 0 & 1 \\ 1 & 0 \end{bmatrix}, \quad \sigma^2 = \begin{bmatrix} 0 & -i \\ i & 0 \end{bmatrix}, \quad \sigma^3 = \begin{bmatrix} 1 & 0 \\ 0 & -1 \end{bmatrix}$$

$$\sigma^m = (\sigma^m_{\alpha\dot\alpha}) \quad , \quad \bar\sigma^m = (\bar\sigma^{m\dot\alpha\alpha})$$

$$\bar\sigma^{m\alpha\dot\alpha} = \varepsilon^{\dot\alpha\dot\beta}\,\varepsilon^{\alpha\beta}\,\sigma^m_{\beta\dot\beta}$$

$$\bar\sigma^0 = \sigma^0 \quad , \quad \bar\sigma^{1,2,3} = -\sigma^{1,2,3}$$

$$\mathrm{Tr}\,\sigma^m\bar\sigma^n \equiv \sigma^m_{\alpha\dot\alpha}\,\bar\sigma^{n\dot\alpha\alpha} = -2\,\eta^{mn}$$

$$\sigma^m_{\alpha\dot\alpha}\,\bar\sigma_m^{\dot\beta\beta} = -2\,\delta_\alpha{}^\beta\,\delta_{\dot\alpha}{}^{\dot\beta}$$

$$(\sigma_{mn})_\alpha{}^\beta = \frac{1}{4}\left(\sigma_m\bar\sigma_n - \sigma_n\bar\sigma_m\right)_\alpha{}^\beta$$

$$(\bar\sigma_{mn})^{\dot\alpha}{}_{\dot\beta} = \frac{1}{4}\left(\bar\sigma_m\sigma_n - \bar\sigma_n\sigma_m\right)^{\dot\alpha}{}_{\dot\beta}$$

$$(\sigma_m\bar\sigma_n)_\alpha{}^\beta = -\eta_{mn}\,\delta_\alpha{}^\beta + 2\,(\sigma_{mn})_\alpha{}^\beta$$

$$(\bar\sigma_m\sigma_n)^{\dot\alpha}{}_{\dot\beta} = -\eta_{mn}\,\delta^{\dot\alpha}{}_{\dot\beta} + 2\,(\bar\sigma_{mn})^{\dot\alpha}{}_{\dot\beta}$$

$$(\sigma_{mn})_\alpha{}^\beta\,\varepsilon_{\beta\gamma} = (\sigma_{mn})_\gamma{}^\beta\,\varepsilon_{\beta\alpha}$$

ε-tensor:

$$(\varepsilon_{\alpha\beta}) = \begin{bmatrix} 0 & -1 \\ 1 & 0 \end{bmatrix} = (\varepsilon_{\dot\alpha\dot\beta}) \quad , \quad (\varepsilon^{\alpha\beta}) = \begin{bmatrix} 0 & 1 \\ -1 & 0 \end{bmatrix} = (\varepsilon^{\dot\alpha\dot\beta})$$

$$\varepsilon_{\alpha\beta}\,\varepsilon^{\beta\gamma} = \delta_\alpha{}^\gamma = \varepsilon^{\gamma\beta}\,\varepsilon_{\beta\alpha} \quad , \quad \varepsilon_{\dot\alpha\dot\beta}\,\varepsilon^{\dot\beta\dot\gamma} = \delta_{\dot\alpha}{}^{\dot\gamma} = \varepsilon^{\dot\gamma\dot\beta}\,\varepsilon_{\dot\beta\dot\alpha}$$

Weyl 2-spinors:

$$\psi'_\alpha = A_\alpha{}^\beta \psi_\beta \qquad \bar\psi_{\dot\alpha} = (A^*)_{\dot\alpha}{}^{\dot\beta} \bar\psi_{\dot\beta}$$

$$\psi'^\alpha = \psi^\beta (A^{-1})_\beta{}^\alpha \qquad \bar\psi'^{\dot\alpha} = \bar\psi^{\dot\beta} (A^{*-1})_{\dot\beta}{}^{\dot\alpha}$$

Raising and lowering of spinorial indices:

$$\psi_\alpha = \varepsilon_{\alpha\beta} \psi^\beta \qquad \psi^\alpha = \varepsilon^{\alpha\beta} \psi_\beta$$

$$\bar\psi_{\dot\alpha} = \varepsilon_{\dot\alpha\dot\beta} \bar\psi^{\dot\beta} \qquad \bar\psi^{\dot\alpha} = \varepsilon^{\dot\alpha\dot\beta} \bar\psi_{\dot\beta}$$

Summation conventions:

$$\zeta\eta = \zeta^\alpha \eta_\alpha \qquad \bar\zeta\bar\eta = \bar\zeta_{\dot\alpha} \bar\eta^{\dot\alpha}$$

Superindices: ("Late-early index convention")

Anholonomic ("flat") : $A \sim (a, \underline{\alpha})$ with $\underline{\alpha} = \alpha$ or $\dot\alpha$

Holonomic ("curved") : $M \sim (m, \underline{\mu})$ with $\underline{\mu} = \mu$ or $\dot\mu$

where $a, b, c, \ldots, m, n, \ell, \ldots \in \{0, 1, 2, 3\}$

$\alpha, \beta, \gamma, \ldots, \mu, \nu, \lambda, \ldots \in \{1, 2\}$

$\dot\alpha, \dot\beta, \dot\gamma, \ldots, \dot\mu, \dot\nu, \dot\lambda, \ldots \in \{1, 2\}$

Graded commutator:

$$[D_A, D_B] = D_A D_B - (-)^{ab} D_B D_A$$

Complex conjugation: Appendix 2

Majorana 4-spinors: Chapter II.7.

APPENDIX 2 : COMPLEX (AND HERMITEAN) CONJUGATION IN SIMPLE SUPERSYMMETRY

Complex conjugation in supersymmetry is a touchy business because of the involved anticommuting variables which transform under different representations of SL(2,ℂ) ; a forteriori this applies to the hermitean conjugation of matrix-valued expressions. (For instance, think about taking the hermitean conjugate of the YM field strength $\mathcal{W}_\alpha = -\frac{1}{8}\bar{D}^2 e^{-V} D_\alpha e^V$.) We will now present a consistent set of rules for these operations. Complex conjugation will be denoted by "*", hermitean conjugation by "+" and all the considered fields may eventually be matrix-valued.

Hermitean (as well as complex) conjugation represents an involutive anti-homomorphism, i.e. the operation is antilinear,

$$(\lambda\phi + \mu\psi)^\dagger = \phi^\dagger\lambda^* + \psi^\dagger\mu^* \qquad (\lambda,\mu \in \mathbb{C}) , \tag{A.1}$$

it reverses the order of all products (including exterior products),

$$(\phi\cdot\psi)^\dagger = \psi^\dagger\cdot\phi^\dagger \tag{A.2}$$

and the composition of the operation with itself reproduces the identity transformation. Its action on tensors and spinors has to be defined in a way which is compatible with their transformation properties.

Weyl 2-spinors with lower dotted, respectively undotted indices transform under SL(2,ℂ) transformations as

$$\psi'_\alpha = A_\alpha{}^\beta\psi_\beta \qquad \text{resp.} \qquad \bar{\psi}'_{\dot\alpha} = (A^*)_{\dot\alpha}{}^{\dot\beta}\,\bar{\psi}_{\dot\beta} \quad . \tag{A.3}$$

These rules are consistent with the definition

$$(\psi_\alpha)^\dagger = \bar{\psi}_{\dot\alpha} \quad . \tag{A.4a}$$

The antisymmetric tensors $(\varepsilon_{\alpha\beta})$, $(\varepsilon_{\dot\alpha\dot\beta})$ are both given by the real matrix

$$\begin{bmatrix} 0 & -1 \\ 1 & 0 \end{bmatrix}$$

and therefore we define

$$\varepsilon_{\alpha\beta}{}^* = \varepsilon_{\dot\alpha\dot\beta} \tag{A.5a}$$

(Similarly for the components of the inverse matrices:

$$(\varepsilon^{\alpha\beta})^* = \varepsilon^{\dot\alpha\dot\beta} \quad .) \tag{A.5b}$$

From the previous equations we have

$$(\psi^{\alpha})^{\dagger} = (\varepsilon^{\alpha\beta}\psi_{\beta})^{\dagger} = \psi_{\beta}^{\dagger}\,(\varepsilon^{\alpha\beta})^* = \bar\psi_{\dot\beta}\,\varepsilon^{\dot\alpha\dot\beta} = \bar\psi^{\dot\alpha} \tag{A.4b}$$

which is consistent with the usual SL(2,$\mathbb{C}$)-transformation properties of ψ^{α} and $\bar\psi^{\dot\alpha}$.

The σ-matrices

$$\sigma^0 = \begin{bmatrix} -1 & \\ & -1 \end{bmatrix} \;, \quad \sigma^1 = \begin{bmatrix} & 1 \\ 1 & \end{bmatrix} \;, \quad \sigma^2 = \begin{bmatrix} & -i \\ i & \end{bmatrix} \;, \quad \sigma^3 = \begin{bmatrix} 1 & \\ & 1 \end{bmatrix}$$

are hermitean, whence the relations

$$(\sigma^a_{\alpha\dot\beta})^* = \sigma^a_{\beta\dot\alpha} \quad , \quad (\bar\sigma^a_{\dot\alpha\beta})^* = \bar\sigma^a_{\dot\beta\alpha} \tag{A.6a}$$

and

$$\left[(\sigma^{ab})_{\alpha}{}^{\beta}\right]^* = -\,(\bar\sigma^{ab})^{\dot\beta}{}_{\dot\alpha} \tag{A.6b}$$

The differentials dz^M have the same reality properties as the coordinates z^M:

$$(dx^m)^* = dx^m \quad , \quad (d\theta^{\alpha})^* = d\bar\theta^{\dot\alpha} \quad , \quad (d\bar\theta_{\dot\alpha})^* = d\theta_{\alpha} \quad . \tag{A.7}$$

For the partial derivatives we define

$$(\partial_m)^* = \partial_m \tag{A.8a}$$

and

$$(\partial_{\alpha})^* = \bar\partial_{\dot\alpha} \qquad \text{i.e.} \qquad \left(\frac{\partial}{\partial\theta^{\alpha}}\right)^* = -\frac{\partial}{\partial\bar\theta^{\dot\alpha}} \quad . \tag{A.8b}$$

By the above rules these relations imply that the invariant derivatives D_{α}, $\bar D_{\dot\alpha}$ are related by complex conjugation:

$$(D_{\alpha})^* = \left(\frac{\partial}{\partial\theta^{\alpha}} + i\,\sigma^m_{\alpha\dot\beta}\,\bar\theta^{\dot\beta}\,\partial_m\right) = \left(\frac{\partial}{\partial\theta^{\alpha}}\right)^* - i\,\partial_m\,(\bar\theta^{\dot\beta})^*\,(\sigma^m_{\alpha\dot\beta})^*$$

$$= -\frac{\partial}{\partial\bar\theta^{\dot\alpha}} - i\,\theta^{\beta}\,\sigma^m_{\beta\dot\alpha}\,\partial_m = \bar D_{\dot\alpha} \quad .$$

It also follows that

$$\left(-\frac{\partial}{\partial\theta_\alpha}\right)^* = (\partial^\alpha)^* = (\varepsilon^{\alpha\beta}\partial_\beta)^* = \varepsilon^{\dot\alpha\dot\beta}\,\bar\partial_{\dot\beta} = \bar\partial^{\dot\alpha} = \frac{\partial}{\partial\bar\theta_{\dot\alpha}} \quad . \tag{A.8c}$$

For a scalar φ and a spinor ψ_β we set

$$(\partial_\alpha\varphi)^\dagger = \varphi^\dagger\,\overleftarrow{\partial}{}^*_\alpha = \partial^*_\alpha\,\varphi^\dagger \tag{A.9a}$$

$$(\partial_\alpha\psi_\beta)^\dagger = \psi_\beta^\dagger\,\overleftarrow{\partial}{}^*_\alpha = -\,\partial^*_\alpha\,\psi_\beta^\dagger \quad . \tag{A.9b}$$

(The last equality is motivated by the change of sign which occurs when fermionic variables are permuted.)

As a general principle for expressions involving several derivatives and several (possibly matrix-valued) scalar or spinorial fields we first completely reverse the order while taking the complex conjugate of all quantities involved. (All derivatives are then assumed to act to their left.) Next the rules (A.9) are systematically applied to the far left of the expression until all derivatives act again to the right. For instance:

$$\begin{aligned}(D^2\varphi)^\dagger = (D^\alpha D_\alpha\varphi)^\dagger = \underbrace{\varphi^\dagger\,\overleftarrow{\bar D}_{\dot\alpha}}\,\overleftarrow{\bar D}{}^{\dot\alpha} &= (\bar D_{\dot\alpha}\varphi^\dagger)\,\overleftarrow{\bar D}{}^{\dot\alpha} \\ &= -\,\bar D^{\dot\alpha}(\bar D_{\dot\alpha}\varphi^\dagger) = \bar D^2\varphi^\dagger\end{aligned}$$

To illustrate the given rules we consider three examples which are of interest in SYM-theories.

(i) The abelian field strength in the chiral basis ($V^\dagger = V$):

$$\begin{aligned}(\bar D^2 D_\beta V)^\dagger = (\bar D_{\dot\alpha}\bar D^{\dot\alpha} D_\beta V)^\dagger = \underbrace{V\,\overleftarrow{\bar D}_{\dot\beta}}\,\overleftarrow{D}{}^\alpha\,\overleftarrow{D}_\alpha &= (\bar D_{\dot\beta}V)\,\overleftarrow{D}{}^\alpha\,\overleftarrow{D}_\alpha \\ = -\,(D^\alpha\bar D_{\dot\beta}V)\,\overleftarrow{D}_\alpha &= -\,D_\alpha D^\alpha\bar D_{\dot\beta}V = D^2\bar D_{\dot\beta}V\end{aligned}$$

(ii) The explicit solution of the constraint $\mathcal{F}_{\alpha\beta} = 0$: ($V = V^\dagger$)

$$(e^{-V}D_\beta e^V)^\dagger = \underbrace{e^V\,\overleftarrow{\bar D}_{\dot\beta}}\,e^{-V} = (\bar D_{\dot\beta}e^V)\,e^{-V} = -\,e^V(\bar D_{\dot\beta}e^{-V})$$

(iii) The vectorial part of the "chiral connection superform":

$$(\bar D_{\dot\alpha}\,\bar\sigma_a^{\dot\alpha\beta}\varphi_\beta)^\dagger = \varphi_\beta^\dagger\,\bar\sigma_a^{\dot\beta\alpha}\,\overleftarrow{D}_\alpha = -\,D_\alpha\,\bar\sigma_a^{\dot\beta\alpha}\,\varphi_\beta^\dagger = -\,D^\alpha(\sigma_a)_{\alpha\dot\beta}(\varphi^\beta)^\dagger .$$

APPENDIX 3 : COMPLEX CONJUGATION IN N=2 SUPERSYMMETRY

Let $(V^i)_{i=1,2}$ be a complex 2-vector transforming under the fundamental representation of SU(2). This representation (denoted by 2) and its complex conjugate (denoted by $\bar{2}$) are equivalent, the intertwiner being given by the ε-matrix:

$$(\varepsilon^{ij}) = \begin{bmatrix} 0 & -1 \\ 1 & 0 \end{bmatrix} \quad , \quad \varepsilon^2 = -\mathbb{1}_2$$

We characterize the vectors transforming under $\bar{2}$ by a lower index, V_i, and we define

$$(V_i)^* = \bar{V}^i \quad . \tag{A.10a}$$

It then follows that

$$\begin{aligned}(V^i)^* &= (\varepsilon^{ij} V_j)^* = (V_j)^* \varepsilon^{ij} = \varepsilon^{ij} \bar{V}^j \\ &= \varepsilon^{ij} \varepsilon^{jk} \bar{V}_k = -\delta^{ik} \bar{V}_k\end{aligned}$$

i.e.

$$(V^i)^* = -\bar{V}_i \quad . \tag{A.10b}$$

With these conventions the fermionic coordinates of U(2)-superspace are related by

$$(\theta^\mu_i)^* = \bar{\theta}^{\dot{\mu} i} \quad .$$

APPENDIX 4 : GEOMETRIC INTERPRETATION OF THE CANONICAL LINEAR CONNECTION ON REDUCTIVE HOMOGENEOUS SPACES

We summarize the results[85] and briefly comment on their implications for rigid superspace $M = SP_o/L_o$.

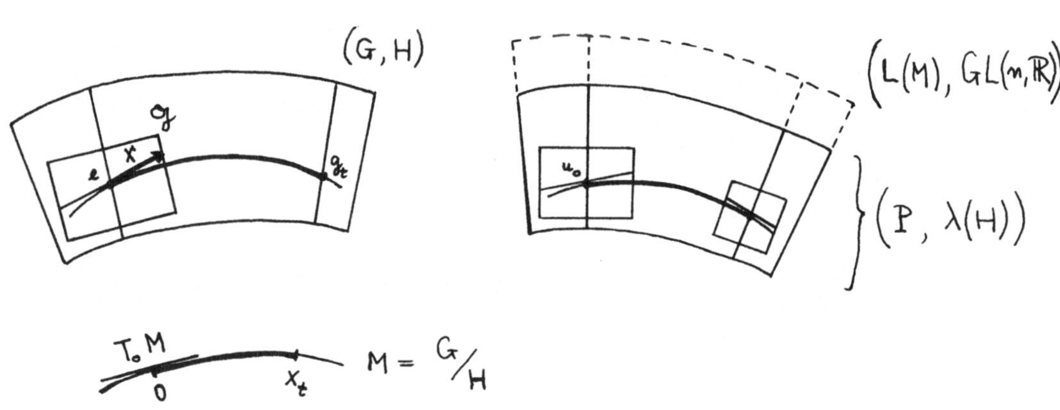

(i) For $X \in \mathfrak{m}$ let $g_t = \exp(tX)$ be the one-parameter subgroup of G generated by X and let $x_t = f_{g_t}(0)$ denote the curve through $0 \in M$ which is induced by the left action of g_t on G. Then the parallel displacement (defined with the canonical linear connection) of tangent vectors at 0 along the curve x_t coincides with the projection

$$\left(f_{g_t}\right)_* \; : \; T_oM \longrightarrow T_{x_t}M$$

(ii) For each $X \in \mathfrak{m}$ the curve x_t defined in (i) is a geodesic with respect to the canonical connection, i.e. $\nabla_{\dot{x}_t}\, \dot{x}_t = 0$ for all t. (Here ∇ denotes the covariant derivative and $\dot{x}_t$ represents the tangent vector to the curve x_t.) Conversely every geodesic starting from $0 \in M$ is of the form $f_{g_t}(0)$ for some $X \in \mathfrak{m}$.

(iii) If T_oM is identified with the vector space $\mathbb{R}^n$ (with the frame u_o) and with $\mathfrak{m}$,

$$\mathbb{R}^n \cong T_oM \cong \mathfrak{m} \quad ,$$

the torsion tensor T and the curvature tensor R of the canonical connection in L(M) can be expressed at $0 \in M$ as

$$T(X,Y)_0 = -[X,Y]_{\mathfrak{m}} \qquad (X,Y \in \mathfrak{m} \cong T_0M)$$

$$(R(X,Y)Z)_0 = -[[X,Y]_h, Z] \qquad (X,Y,Z \in \mathfrak{m} \cong T_0M).$$

Here we used the notation

$$[X,Y] = [X,Y]_h + [X,Y]_{\mathfrak{m}}$$

with

$$[X,Y]_h \in h \quad , \quad [X,Y]_{\mathfrak{m}} \in \mathfrak{m}$$

for $X,Y \in \mathfrak{g} = h + \mathfrak{m}$.

For the super-Poincaré group we have $h = \mathrm{Lie}\, L_o$ and $[X,Y]_h = 0$ for $X,Y \in \mathfrak{m}$, hence

$$(R(X,Y)Z)_0 = 0 \qquad \text{for all} \qquad X,Y,Z \in \mathfrak{m}.$$

APPENDIX 5 : KOSZUL'S FORMULA ("BRS COHOMOLOGY")

The aim of this appendix is to establish an alternative formula for the coboundary operator of the Lie algebra cohomology (eq. (4.30a)). As in chapter IV.1 (vii) we consider a finite dimensional Lie group G with Lie algebra $\mathfrak{g}$ and a linear representation t of $\mathfrak{g}$ on a finite dimensional vector space V.

Let $\{e_i\}_{i=1,\ldots,\dim G}$ be a basis of $\mathfrak{g}$ with structure constants $f_{ij}{}^k$,

$$[e_i , e_j] = f_{ij}{}^k e_k$$

and let $\{t_i\}$ denote its representation on V :

$$t : \mathfrak{g} \longrightarrow \mathcal{E}nd(V)$$

$$e_i \longmapsto t(e_i) \equiv t_i$$

$$[t_i , t_j] = f_{ij}{}^k t_k . \tag{A.11}$$

Now consider $\mathfrak{g}^*$, the vector space dual of $\mathfrak{g}$ and the dual basis $\{e^i\}$ of $\{e_i\}$. On the direct sum of these vector spaces, $\mathfrak{g}^* + \mathfrak{g}$, we define an inner product (,) by

$$\begin{aligned} (e^i , e^j) &= 0 = (e_i , e_j) \\ (e^i , e_j) &= (e_j , e^i) = \delta^i{}_j . \end{aligned} \tag{A.12a}$$

To proceed further, we consider the Clifford algebra[189)] associated to the vector space $\mathfrak{g}^* + \mathfrak{g}$ and to the inner product (A.12a):

$$j : \mathfrak{g}^* + \mathfrak{g} \xrightarrow{\text{linear}} \text{Clifford algebra}$$

$$e^i \longmapsto c^i$$

$$e_i \longmapsto b_i$$

$$\begin{aligned} \{c^i , c^j\} &= 0 = \{b_i , b_j\} \\ \{c^i , b_j\} &= \delta^i{}_j . \end{aligned} \tag{A.12b}$$

186)

An appropriate representation space has to be chosen for this algebra. For this purpose we regard c^i and b_i, respectively, as creation and annihilation operators acting on a Fock space $\mathcal{F}$ defined by a vacuum state $|0\rangle$: the vacuum is annihilated by the operators b_i,

$$b_i\,|0\rangle = 0 \qquad \text{for} \qquad i \in \{1, \ldots, \dim G\}$$

and the other states are constructed by having the creation operators act on $|0\rangle$.

Finally we provide $\mathcal{F}$ with a grading (ghost number, BRS charge):

$$\deg|0\rangle = 0 \quad , \quad \deg c^i = 1 = -\deg b_i \quad .$$

Thus we have a direct sum decomposition of the representation spaces $\mathcal{F}$ and $V \otimes \mathcal{F}$:

$$\mathcal{F} = \bigoplus_{i=1}^{\dim G} \mathcal{F}^p$$

$$V \otimes \mathcal{F} = \bigoplus_{i=1}^{\dim G} \mathcal{C}^p \qquad (\deg V = 0) \quad .$$

The elements of $\mathcal{C}^p$ (i.e. the V-valued states with ghost number p) are equivalent to the previously defined p-cochains with values in V.

The coboundary operator δ is realized on $V \otimes \mathcal{F}$ by

$$Q = t_i \otimes c^i + 1 \otimes \left[-\tfrac{1}{2} f_{ij}{}^k\, c^i c^j b_k\right] \tag{A.13}$$

$$\equiv c^i t_i - \frac{1}{2} f_{ij}{}^k\, c^i c^j b_k \quad .$$

This can be seen as follows. To the 2-cochain

$$\omega : \mathfrak{g} \times \mathfrak{g} \longrightarrow V$$

$$(X, Y) \longmapsto \omega(X,Y) = \omega(X^i e_i , Y^j e_j)$$

$$= \omega(e_i, e_j)\, X^i Y^j \equiv \omega_{ij} X^i Y^j$$

we associate the antisymmetric form $\omega_{ij} c^i c^j$ $(\omega_{ij} \in V)$; a short calculation then shows that

$$[Q, \omega_{ij} c^i c^j] \equiv c^k t_k \omega_{ij} c^i c^j - [\tfrac{1}{2} f_{ij}{}^k c^i c^j b_k , \omega_{\ell m} c^\ell c^m]$$
$$= c^i c^j c^k \{ t_i \omega_{jk} - f_{ij}{}^m \omega_{mk} \}$$

which result is in agreement with

$$(\delta\omega)(X, Y, Z) = \tfrac{1}{3} X^i Y^j Z^k \{ t_i \omega_{jk} - f_{ij}{}^m \omega_{mk} \}.$$

Schematically:

$$\begin{array}{ccc} \omega(X,Y) \equiv \omega_{ij} X^i Y^j & \longmapsto & \omega_{ij} c^i c^j \\ \downarrow \delta & & \downarrow Q \\ (\delta\omega)(X,Y,Z) \equiv \tfrac{1}{3} \Omega_{ijk} X^i Y^j Z^k & \longmapsto & [Q, \omega_{ij} c^i c^j] = \Omega_{ijk} c^i c^j c^k \end{array}$$

(similarly for arbitrary p-cochains).

The operator Q increases the ghost number by one unit,

$$Q : \mathcal{C}^p \longrightarrow \mathcal{C}^{p+1}$$

and satisfies the nilpotency relation

$$Q^2 = \tfrac{1}{2} \{ Q , Q \} = 0$$

which equation can be directly verified by use of (A.11),(A.12b) and of the Jacobi identity for $\mathfrak{g}$:

$$f_{ij}{}^k f_{k\ell}{}^m + f_{\ell i}{}^k f_{kj}{}^m + f_{j\ell}{}^k f_{ki}{}^m = 0 .$$

Consequently one can define cocycles, coboundaries and cohomology classes for the complex $(V \otimes \mathcal{F}, Q)$ in complete analogy to the complex $(C^*(\mathfrak{g}, V), \delta)$, see eqs (4.30d). In the Physics literature this cohomology is often referred to as BRS-cohomology[187].

In the case where V equals $\mathbb{R}$ (i.e. the field over which the Lie algebra $\mathfrak{g}$ is defined) the representation t on V is trivial ($t_i = 0$) and the first term in (A.13) drops out. This corresponds to the usual form of Koszul's formula[185] as presented e.g. in ref. 63)*).

Concerning the physical interpretation and application of the presented formalism we remark the following. In the Lagrangian framework (considered in the body of these notes) the operators c^i correspond to the ghost fields and the b_i are represented by $\delta/\delta c^i$, both acting on the vector space of functionals of c^i and of the classical fields, see 136) section III.4. (The algebraic structure of the antighost fields occuring in this context is not determined by the underlying symmetry algebra, rather it depends on the choice of the gauge fixing functions.) On the other hand, in the Hamiltonian approach[190] the b_i describe the "antighosts" (which do not coincide with those of the Lagrangian framework).

The mathematical formalism introduced above can be generalized in a straightforward way[186],[187],[188] to the case of infinite dimensional Lie groups and algebras like the Virasoro algebra occuring in string theories[173]. However, in this context it is necessary to introduce a normal ordering procedure for the multiplication of operators with the result that the operator Q is only nilpotent under specific conditions. (For details see the references cited.)

*) For a direct comparison with 63) consider the following definition of the Lie derivative as acting on p-cochains ω :

$$(L_X \omega)(Y_1, \ldots, Y_p) = \sum_{i=1}^{p} \omega(Y_1, \ldots, [Y_i, X], \ldots, Y_p)$$

APPENDIX 6 : ON THE DESCRIPTION OF ANTICOMMUTING SPINORS IN ORDINARY AND SUPERSYMMETRIC FIELD THEORIES

In the following we present an elementary method for describing 'classical' anticommuting Fermi fields and functionals of these variables. This approach (which straightforwardly extends to supersymmetric theories) avoids the *ab initio* introduction of infinite dimensional Grassmann algebras.

(i) Introduction

The path integral approach to the quantization of fermionic fields only leads to consistent results, if the ('classical') fields are taken to be anticommuting variables. This assumption also applies to the spinorial components of the superfunctions in supersymmetric field theories. Obviously the occurrence of these classical anticommuting fields in ordinary and supersymmetric field theories requires the introduction of infinite dimensional Grassmann algebras. The infinite dimensionality is due to the space-time continuum and therefore these algebras should not be introduced without any reference to this structure*), rather they should be generated by the space-time geometry.

In the following we will present a simple implementation of this idea and show how functionals of spinor fields can be given a well-defined meaning in these terms**). Fermi fields in d-dimensional Minkowski space will be viewed as (improper) generators of the antisymmetric tensor algebra over $\mathcal{S}'(\mathbb{R}^d)\otimes \mathbb{C}^N$. (Here $\mathcal{S}'(\mathbb{R}^d)$ denotes the space of tempered distributions on $\mathbb{R}^d$ and N is the dimension of the associated Dirac algebra.) Functionals of these fields can be expanded with respect to the generators. More specifically our formalism leads to anticommuting component fields in the superfunctions of Berezin-Leĭtes[195),196)] and Kostant-Batchelor[88),197)] and thereby solves an open problem[117)] in the physical application[180),214)] of this elegant and powerful theory.

It should be emphasized that our description is primarily aimed at a conceptual clarification rather than any computational applications. Also we should briefly mention the history of this classical subject. Anticommuting sources

---

*) As done in several mathematical approaches to supersymmetry[116),82),194)]

**) Our interpretation has already been alluded to in some earlier unpublished notes of R. Stora[182)].

have been introduced in 1951 in the pioneering work of J. Schwinger[199]. Thereafter functional integrals over such fields have been formally defined and evaluated by P.T. Mathews and A. Salam[200]. Much of the subsequent work has been aimed at promoting to a rigorous level the original formal concepts and manipulations. We quote the classical study of F.A. Berezin[201] where various new concepts have been introduced and the Euclidean space approach of Osterwalder and Schrader[202] which involves some novel features even at a purely formal level. Recently many authors have tried to develop a satisfactory theory of integration over Grassmann variables and supermanifolds which mimicks the operational Berezin integration rules[195),196),203),214]. Finally a rigorous analytic scheme for fermionic path integration has been proposed in a very recent article[198] where further references on these topics can be found.

This appendix is organized as follows. After illustrating the basic idea in the case of a scalar field, we discuss spinor fields and compare our approach with the one of Berezin. Subsequently these ideas are extended to superfields and generalized from d-dimensional Minkowski space to more general space-time manifolds.

(ii) Scalar fields

To start with we introduce some notation and illustrate our approach by considering a single neutral scalar field φ on $\mathbb{R}^d$. Although the usual definitions for the action and vertex functionals are well-defined in this case[174] they can be reinterpreted and formulated in a way which generalizes in a straightforward way to the fermionic theory.

The scalar field φ may be viewed as an element of the Schwartz space of strongly decreasing test-functions, $\mathcal{S}(\mathbb{R}^d)$. The action and the vertex functional are formal functionals of φ which are determined by a sequence of symmetric distribution-valued integral kernels $T_n(x_1,\ldots,x_n)$:

$$\mathcal{A}[\varphi] = \sum_{n=0}^{\infty} \frac{1}{n!} \int dx_1 \ldots \int dx_n \, T_n(x_1,\ldots,x_n)\, \varphi(x_1)\cdot\ldots\cdot\varphi(x_n) \tag{A.14}$$

For instance the action of the φ^4-theory is given by

$$\mathcal{A}[\varphi] = \int dx \left[\frac{1}{2} (\partial_\mu \varphi)(\partial^\mu \varphi) - \frac{m^2}{2} \varphi^2 - \frac{\lambda}{4!} \varphi^4 \right]$$

$$= \int dx_1 \int dx_2 \left[-\frac{1}{2} \delta(x_1 - x_2) \left(\Box_{x_1} + m^2 \right) \right] \varphi(x_1)\, \varphi(x_2)$$

$$+ \int dx_1 \ldots \int dx_4 \left[\frac{-\lambda}{4!} \delta(x_1 - x_2)\,\delta(x_2 - x_3)\,\delta(x_3 - x_4) \right] \varphi(x_1) \cdot \ldots \cdot \varphi(x_4) \,.$$

In slightly different words: <u>The functional $A[\cdot]$ is specified by a sequence of distributions</u> $T_n(x_1,\ldots,x_n)$,

$$T = \begin{bmatrix} 1 \\ T_1 \\ T_2 \\ \vdots \end{bmatrix} \qquad \longleftrightarrow \qquad A[\cdot] \tag{A.15}$$

and the action $A[\varphi]$ is determined by acting formally with T on the following sequence of test functions:

$$\phi = \begin{bmatrix} 1 \\ \phi_1 \\ \phi_2 \\ \vdots \end{bmatrix} \,, \quad \phi_n(x_1,\ldots,x_n) \equiv (\underbrace{\varphi \otimes \ldots \otimes \varphi}_{\text{n factors}})(x_1,\ldots,x_n) = \varphi(x_1) \cdot \ldots \cdot \varphi(x_n) \tag{A.16}$$

Explicitly:

$$T[\phi] = \sum_{n=0}^{\infty} T_n[\phi_n] = \sum_{n=0}^{\infty} \frac{1}{n!} \int dx_1 \ldots \int dx_n \, T_n(x_1,\ldots,x_n)\, \varphi(x_1) \cdot \ldots \cdot \varphi(x_n)$$

$$= A[\varphi]$$

In the previous equations we assumed that $x_i \in \mathbb{R}^d$ and we used the notation

$$\delta(x-y) \equiv \delta^{(d)}(x-y) \equiv \delta(x^1 - y^1) \cdot \ldots \cdot \delta(x^d - y^d) \qquad (x, y \in \mathbb{R}^d).$$

Furthermore the action of distributions $T_n \in \mathcal{S}'(\mathbb{R}^{n.d})$ on test functions $\phi_n \in \mathcal{S}(\mathbb{R}^{n.d})$ has been written symbolically as

$$T_n[\phi_n] = \frac{1}{n!} \int dx_1 \ldots \int dx_n \, T_n(x_1,\ldots,x_n)\, \phi_n(x_1,\ldots,x_n) \,. \tag{A.17}$$

In general $T_n(x_1,\ldots,x_n)$ does not represent an ordinary function, but just a formal device[204].

The notation (A.17) corresponds to the definition

$$T_n = \frac{1}{n!}\int dx_1 \ldots \int dx_n \; T_n(x_1,\ldots,x_n)\; \delta_{x_1} \otimes \ldots \otimes \delta_{x_n} \qquad \text{(A.18)}$$

where $\delta_{x_1} \otimes \ldots \otimes \delta_{x_n}$ is the n-fold tensor product of Dirac δ-distributions with support in $x_1,\ldots,x_n$,

$$\delta_{x_1}[F_1] = F_1(x_1) \qquad F_1 \in \mathcal{S}(\mathbb{R}^d)$$

$$\vdots$$

$$(\delta_{x_1} \otimes \ldots \otimes \delta_{x_n})[F_n] = F_n(x_1,\ldots,x_n) \qquad F_n \in \mathcal{S}(\mathbb{R}^{n\cdot d}).$$

For instance the δ-functional with support in y is given by

$$\delta_y = \int dx \; \delta(x-y)\, \delta_x$$

since

$$\delta_y[F_1] = \int dx \; \delta(x-y)\, \delta_x[F_1]$$

$$= \int dx \; \delta(x-y)\, F_1(x) = F_1(y) .$$

Accordingly the δ-functionals can be interpreted as an improper basis of the space of tempered distributions and every element $T_n \in \mathcal{S}'(\mathbb{R}^{n.d})$ can be formally expanded w.r.t. these generators as in eq. (A.18)*).

While eq. (A.14) is formally defined for functions $\varphi(x)$, it does not make sense any more if we pass over to spinor fields with anticommuting components $\Psi_i(x)$. Therefore we look for an equivalent expression for (A.14) which still has a well-defined meaning in the fermionic case.

To formulate our results in standard mathematical terms we introduce the <u>tensor algebra over $\mathcal{S}(\mathbb{R}^d)$</u>:

$$\mathcal{S} = \bigoplus_{n=0}^{\infty} \mathcal{S}(\mathbb{R}^{n\cdot d})$$

---

*) Of course there is a similar expansion for the test functions, e.g. for $F_1 \in \mathcal{S}(\mathbb{R}^d)$:

$$F_1 = \int dx \; F_1(x)\, \delta_x \qquad i.e. \qquad F_1(y) = \int dx \; F_1(x)\, \delta(x-y).$$

where

$$\mathcal{S}(\mathbb{R}^0) \equiv \mathbb{C} \;, \quad \mathcal{S}(\mathbb{R}^{n\cdot d}) \cong \mathcal{S}(\mathbb{R}^d) \otimes \ldots \otimes \mathcal{S}(\mathbb{R}^d) .$$

By definition[205] $\mathcal{S}$ contains sequences of the form $F = (F_0, F_1, \ldots)$ with $F_n \in \mathcal{S}(\mathbb{R}^{n.d})$ and only a finite number of elements $F_n \neq 0$. (Strictly speaking the sequence (A.16) does not belong to the algebra $\mathcal{S}$ since it involves infinitely many non-vanishing elements. Thus the evaluation of linear functionals on this sequence may lead to convergence problems, see below.)

The space $\mathcal{S}$ supplemented with the direct sum topology[205),206] is a topological vector space and therefore one can consider continuous linear functionals on $\mathcal{S}$. (A.15) represents such a functional and belongs to the dual $\mathcal{S}^*$ of $\mathcal{S}$ which can be algebraically identified with the product space

$$\bigotimes_{n=0}^{\infty} \mathcal{S}'(\mathbb{R}^{n\cdot d}) \quad , \qquad \mathcal{S}'(\mathbb{R}^0) \equiv \mathbb{C}$$

consisting of sequences $T=(T_0, T_1, \ldots)$ of tempered distributions: $T_n \in \mathcal{S}'(\mathbb{R}^{n.d})$. (The identification is defined by the relation

$$T[F] = \sum_{n=0}^{\infty} T_n[F_n] \qquad (T \in \mathcal{S}^* , \; F \in \mathcal{S}) ,$$

Again $\mathcal{S}^*$ (supplemented with the strong topology[207]) is a topological vector space.

Since scalar particles obey the Bose-Einstein statistics we restrict our attention to the symmetric test function algebra,

$$\mathcal{S}_{sym} = \bigoplus_{n=0}^{\infty} \mathcal{S}_{sym}(\mathbb{R}^{n\cdot d})$$

$$\mathcal{S}_{sym}(\mathbb{R}^{n\cdot d}) \cong \mathcal{S}(\mathbb{R}^d) \otimes_s \ldots \otimes_s \mathcal{S}(\mathbb{R}^d) \quad \text{(symmetrized tensor product)}$$

and to the corresponding strong dual, $(\mathcal{S}_{sym})^*$. Let $\varphi(x)$ denote an <u>(improper) generator in degree one of $(\mathcal{S}_{sym})^*$</u>. Its action on an element

$$F = \begin{bmatrix} 1 \\ F_1 \\ F_2 \\ \vdots \end{bmatrix} \in \mathcal{J} \tag{A.19}$$

is given by

$$\begin{aligned} \varphi(x)[F] &= F_1(x) \\ &\vdots \\ \left(\varphi(x_1) \otimes \ldots \otimes \varphi(x_n)\right)[F] &= F_n(x_1, \ldots, x_n) \end{aligned}$$

where F_n is symmetric in all its arguments. Identifying $(\mathcal{J}_{sym}(\mathbb{R}^{n \cdot d}))^*$ with its image under the canonical injection

$$\begin{aligned} i_n : \mathcal{J}_{sym}(\mathbb{R}^{n \cdot d})^* &\hookrightarrow \mathcal{J}^*_{sym} \\ T_n &\longmapsto (0, \ldots, 0, T_n, 0, \ldots) \end{aligned} \tag{A.20}$$

we can write

$$\mathcal{J}^*_{sym} = \sum_{n=0}^{\infty} \mathcal{J}_{sym}(\mathbb{R}^{n \cdot d})^* . \tag{A.21}$$

A <u>general element $T_s \in \mathcal{J}^*_{sym}$</u> then has the form

$$T_s = \sum_{n=0}^{\infty} \frac{1}{n!} \int dx_1 \ldots \int dx_n \, T_n(x_1, \ldots, x_n) \, \varphi(x_1) \otimes \ldots \otimes \varphi(x_n) \tag{A.22}$$

and application to the test functions (A.19) yields

$$T_s[F] = \sum_{n=0}^{\infty} \frac{1}{n!} \int dx_1 \ldots \int dx_n \, T_n(x_1, \ldots, x_n) \, F_n(x_1, \ldots, x_n).$$

In particular, if the sequence $(1, T_1, T_2, \ldots)$ represents the classical action of the scalar field φ and if ϕ denotes the sequence of test functions (A.16), we recover our initial equation (A.14):

$$T_s[\phi] = \sum_{n=0}^{\infty} \frac{1}{n!} \int dx_1 \ldots \int dx_n \, T_n(x_1,\ldots,x_n) \, \varphi(x_1) \cdot \ldots \cdot \varphi(x_n)$$

$$= A[\varphi] \, .$$

(As pointed out before the sequence (A.16) does not belong to the tensor algebra $\mathcal{S}$ and therefore the evaluation of T_s on it may lead to convergence problems.)

Observe that the expansion (A.22) of the symmetric functional T_s w.r.t. the generators $\varphi(x)$ corresponds to a symmetrized version of the expansion of tempered distributions w.r.t. the δ-functionals, eq. (A.18). Actually (A.22) is the equivalent expression we have been looking for: this quantity still has a well-defined meaning, if we replace $\varphi(x)$ by a generator $\underline{\Psi}_i(x)$ of the antisymmetric tensor algebra.

(iii) Spinor fields

To take care of the N components of a spinor in d dimensions we define*)

$$\mathcal{S}_{i_1 \ldots i_n} := \mathcal{S}(\mathbb{R}^{n \cdot d}) \quad \text{for} \quad i_1, \ldots, i_n \in \{1, \ldots, N\}$$

$$\cong \mathcal{S}(\mathbb{R}^d) \otimes \ldots \otimes \mathcal{S}(\mathbb{R}^d) \quad \text{(n factors)}$$

$$\mathcal{S}_n := \bigoplus_{i_1,\ldots,i_n=1}^{N} \mathcal{S}_{i_1 \ldots i_n} \quad , \quad \mathcal{S}_0 \equiv \mathbb{C}$$

$$\mathcal{S} := \bigoplus_{n=0}^{\infty} \mathcal{S}_n \tag{A.23}$$

$$= \mathbb{C} \oplus \underbrace{\mathcal{S}(\mathbb{R}^d) \oplus \ldots \oplus \mathcal{S}(\mathbb{R}^d)}_{N \text{ terms}} \oplus \underbrace{\mathcal{S}(\mathbb{R}^{2 \cdot d}) \oplus \ldots \oplus \mathcal{S}(\mathbb{R}^{2 \cdot d})}_{N^2 \text{ terms}} \oplus \ldots$$

---

*) This is also the set-up of the Borchers algebra in the axiomatic quantum theory of N fields[288].

Thus a typical element of $\mathcal{S}$ has the form

$$F = (\underbrace{\overset{\in \mathbb{C}}{F_0}}_{\in \mathcal{S}_0} ; \underbrace{\overset{\in \mathcal{S}(\mathbb{R}^d)}{F_1, \ldots, F_N}}_{\in \mathcal{S}_1} ; \underbrace{\overset{\in \mathcal{S}(\mathbb{R}^{2\cdot d})}{F_{11}, F_{21}, \ldots}}_{\in \mathcal{S}_2} ; \ldots ; \underbrace{\ldots, \overset{\in \mathcal{S}(\mathbb{R}^{n\cdot d})}{F_{i_1 \ldots i_n}}, \ldots}_{\in \mathcal{S}_n} ; 0, 0, \ldots) .$$

The algebra $\mathcal{S}_{\text{antisym}}$ of antisymmetric test functions is obtained by taking the antisymmetric tensor product in the previous definition:

$$\left(\mathcal{S}_{i_1 \ldots i_n}\right)_{\text{antisym}} \cong \mathcal{S}(\mathbb{R}^d) \otimes_a \ldots \otimes_a \mathcal{S}(\mathbb{R}^d) .$$

Introducing generators $\underline{\Psi}_1(x), \ldots, \underline{\Psi}_N(x)$ for the tensor algebra of antisymmetric distributions, $(\mathcal{S}_{\text{antisym}})^*$, we can represent the action functional as

$$T_a = \sum_{n=0}^{\infty} \sum_{i_1, \ldots, i_n = 1}^{N} \int dx_1 \ldots \int dx_n \, T_{i_1 \ldots i_n}(x_1, \ldots, x_n) \, \underline{\Psi}_{i_1}(x_1) \otimes \ldots \otimes \underline{\Psi}_{i_n}(x_n) . \tag{A.24}$$

Note that the generators $\underline{\Psi}_i(x)$ satisfy the antisymmetry property

$$\begin{aligned} \left(\underline{\Psi}_i(x_1) \otimes \underline{\Psi}_j(x_2)\right)[F] &= F_{ij}(x_1, x_2) = - F_{ij}(x_2, x_1) \\ &= - \left(\underline{\Psi}_i(x_2) \otimes \underline{\Psi}_j(x_1)\right)[F] . \end{aligned}$$

This is the appropriate moment to point out some relations of our previous definitions with the concept of rigged Hilbert space[209)] which is also at the very heart of Berezin's approach to anticommuting spinor fields (see p. 62 of ref. 201)). Obviously the Hilbert space which lies at the base of our exposition is $\mathcal{L}^2(\mathbb{R}^d, dx)$ "equipped" with the nuclear space $\mathcal{S}(\mathbb{R}^d)$ and its dual:

$$\mathcal{S}(\mathbb{R}^d) \subset \mathcal{L}^2(\mathbb{R}^d, dx) \subset \mathcal{S}'(\mathbb{R}^d) . \tag{A.25}$$

This set-up is well-known from Quantum Mechanics[22),210)]. Any $\mathcal{L}^2$-function g can be viewed as a regular tempered distribution,

$$g : \mathcal{S}(\mathbb{R}^d) \xrightarrow{\text{linear}} \mathbb{C}$$

$$f \longmapsto \int_{\mathbb{R}^d} dx\, g^*(x)\, f(x)$$

and the linear space $\mathcal{S}(\mathbb{R}^d)$ is a common domain of essential self-adjointness for the position operators $(x^k)_{k=1,\dots,d}$. On the configuration space $\mathbb{R}^d$ these observables admit a continuous spectrum of eigenvalues and the corresponding eigen-"functions" $\delta(x-y)$ do not belong to the Hilbert space $\mathcal{L}^2(\mathbb{R}^d, dx)$, rather they define a generalized basis of $\mathcal{S}'(\mathbb{R}^d)$.

The rigged Hilbert space which is underlying our foregoing discussion of spinor fields is given by

$$\bigoplus_{n=0}^{\infty} \mathcal{S}_{n,\,antisym} \subset \mathcal{F}_a \subset \left(\bigoplus_{n=0}^{\infty} \mathcal{S}_{n,\,antisym} \right)^* ,$$

where $\mathcal{F}_a$ denotes the antisymmetric Fock space obtained by inner product completion of

$$\mathbb{C} \oplus \mathcal{H} \oplus (\mathcal{H} \otimes_a \mathcal{H}) \oplus \dots$$

$$\mathcal{H} = \mathcal{L}^2(\mathbb{R}^d, dx) \otimes \mathbb{C}^N .$$

Note that these are also the spaces which are needed for the quantum theory: the Fock space of states, the set of test functions with which the quantum operators have to be smeared and the space of generalized eigenfunctions of these operators.

On his part Berezin (who only considers the case N = 1) starts with a so-called infinite dimensional Grassmann algebra $\mathcal{G} = \bigoplus_{n=0}^{\infty} E^n$ with inner product (,). This algebraic structure is defined in such a way that for each 'subspace' E^n of $\mathcal{G}$ there is an associated rigged Hilbert space

$$\tilde{E}^n \subset \mathcal{H}^n \subset E^n , \quad n \in \{0, 1, 2, \dots\} .$$

A generalized orthonormal basis $\{\underline{\alpha}(x)\}$ for the Hilbert space $\mathcal{H}^1$ is then introduced by means of an isometric mapping α from the $\mathcal{L}^2$-functions on some space M (with measure dx) onto $\mathcal{H}^1$,

$$\alpha : \mathcal{L}^2(\mathbb{R}^d, dx) \xrightarrow{\text{linear}} \mathcal{H}^1$$

$$f \longmapsto \alpha[f] = \int_M dx\ \underline{\alpha}(x)\, f(x)$$

$$(\alpha[f], \alpha[f]) = \int_M dx\, |f(x)|^2 . \tag{A.26}$$

In this approach an arbitrary element $f \in \mathcal{G} = \bigoplus_{n=0}^{\infty} E^n$ admits the representation

$$f = \sum_{n=0}^{\infty} \int_M dx_1 \ldots \int_M dx_n\, f_n(x_1, \ldots, x_n)\, \underline{\alpha}(x_1) \cdot \ldots \cdot \underline{\alpha}(x_n)$$

where f_n is a (generalized) antisymmetric function of n variables. The product of the so-called generators $\underline{\alpha}(x)$ of $\mathcal{G}$ is formally defined by

$$\left(\int dx\, f_1(x)\, \underline{\alpha}(x) \right) \wedge \left(\int dy\, g_1(y)\, \underline{\alpha}(y) \right) =: \int dx \int dy\, f_1(x)\, g_1(y)\, \underline{\alpha}(x) \cdot \underline{\alpha}(y)$$

and thereby satisfies the antisymmetry property

$$\underline{\alpha}(x) \cdot \underline{\alpha}(y) = - \underline{\alpha}(y) \cdot \underline{\alpha}(x) .$$

In the sequel of his exposition Berezin supplements the Grassmann algebra $\mathcal{G}$ with an involution $f \mapsto f^*$ and defines functional integrals over the generators $\underline{\alpha}(x)$, $\underline{\alpha}^*(x)$. The latter allow for a formal derivation of the familiar equations for the determinants *).

Clearly there are many close parallels between these two approaches, but as far as Physics is concerned, we feel that our Ansatz is more natural from a conceptual point of view and also more economical in the sense that it only introduces the spaces that are anyhow needed for the quantization of the theory.

---

*) Since our test function algebra and its dual admit a natural operation of involution[208], these considerations also apply in this case.

(iv) Superfields

Let us first recall a few basic notions. Superfields are usually given by their "θ-expansion",

$$\phi(x,\theta) = C(x) + \theta_\alpha \chi^\alpha(x) + \theta_\alpha \theta_\beta v^{\alpha\beta}(x) + \dots \tag{A.27}$$

where the x-independent and anticommuting variables θ_α generate the Grassmann algebra over some finite-dimensional vector space V:

$$\{\theta_\alpha, \theta_\beta\} = 0 \quad .$$

In the field theoretical applications the dimension of V coincides with the dimension of the Dirac algebra in d-space-time dimensions, e.g. for d = 4:

$$(\theta_\alpha) = (\theta_1, \theta_2, \bar{\theta}^{\dot{1}}, \bar{\theta}^{\dot{2}})^t \quad .$$

The "duals" $d\theta^\alpha$ of the generators θ_α are defined by means of the Berezin integral[201)] which is a purely algebraic operation:

$$\int d\theta^\alpha \, \theta_\beta = \delta^\alpha{}_\beta \quad .$$

The variables $d\theta^\alpha$ also anticommute with each other and should not be confused with the (commuting) differential 1-forms that are misleadingly denoted by the same symbols.

Actions and other functionals of superfields are "integrals over superspace" (or subspaces of it), e.g. for d = 4

$$\int dx \int d^4\theta \, \phi(x,\theta)^2 \tag{A.28}$$

$$= \int dx_1 \int d^4\theta_1 \int dx_2 \int d^4\theta_2 \left[\delta(x_1 - x_2)\, \delta(\theta_1 - \theta_2)\right] \phi(x_1,\theta_1)\, \phi(x_2,\theta_2)$$

$$= \int dx_1 \int d^4\theta_1 \int dx_2 \int d^4\theta_2 \; T_2(x_1,\theta_1; x_2,\theta_2)\, \phi(x_1,\theta_1)\, \phi(x_2,\theta_2).$$

As in the last sections T_2 represents a distribution-valued integral kernel which in the case at hand also depends on the Grassmann variables θ_α . Obviously functionals of superfields correspond to sequences of such integral kernels.

We now reformulate the previous equations in slightly more mathematical terms. Consider a N-dimensional ($N < \infty$) complex vector space V with a basis $\{ e_i \}_{i=1}^{N}$ and the generators $\{ j(e_i) \}_{i=1}^{N}$ of the associated Grassmann algebra G(V)[189]:

$$\{ j(e_i), j(e_k) \} = 0 \quad \text{for} \quad i,k \in \{1,\dots,N\} .$$

In the following we will only deal with the generators $j(e_i)$ (which correspond to the variables θ_α) and therefore no confusion should arise from suppressing the symbol j that denotes the linear homomorphism $j : V \longrightarrow G(V)$.

The basis of G(V) generated by $\{ e_i \}_{i=1}^{N}$ is given by the 2^N monomials

$$1, \; e_{i_1}, \; e_{i_1} \wedge e_{i_2}, \dots, e_{i_1} \wedge \dots \wedge e_{i_m}, \dots, e_1 \wedge \dots \wedge e_N$$

$$(1 \le i_1 < \dots < i_m \le N) \tag{A.29}$$

where the "wedge" represents the multiplication in G(V). To simplify the notation the basis elements (A.29) will be labelled by $\{ e_I \}_{I=1}^{2N}$.

Let V^* be the dual vector space of V. The generators of $G(V^*)$ which are dual to $e_1,\dots,e_N$ are to be characterized by upper indices, $\{ e^i \}_{i=1}^{N}$,

$$e^i(e_j) = \delta^i{}_j \quad \text{for} \quad i,j \in \{1,\dots,N\} ,$$

similarly for the associated basis $\{ e^I \}_{I=1}^{2^N}$ of $G(V^*)$.

In order to reproduce the superfield functionals in the form (A.28) we have to make contact with the Berezinian integral via the relation *)

$$\varepsilon^{i_1\dots i_m}_{j_1\dots j_m} = e^{i_1} \wedge \dots \wedge e^{i_m} (e_{j_1} \wedge \dots \wedge e_{j_m}) \tag{A.30}$$

$$= \int \left(\sum_{H=1}^{2^N} e^H \right) \left(\sum_{l=0}^{N} \sum_{1 \le k_1 < \dots < k_l \le N} \varepsilon^{i_1 \dots i_m}_{k_1 \dots k_l} e_{k_1} \wedge \dots \wedge e_{k_l} \right) \wedge \left(e_{j_1} \wedge \dots \wedge e_{j_m} \right),$$

---

*) Here $\varepsilon^{i_1\dots i_n}_{j_1\dots j_m}$ is zero, if $n \neq m$ or if $n = m$, but $\{i_1,\dots,i_n\} \neq \{j_1,\dots,j_m\}$; it is + 1 (- 1), if both sequences of indices are related by an even (odd) number of transpositions.

the appropriate combinatorial factor being absorbed in the sum over H, e.g.

$$\delta^i_j = e^i(e_j) = \int \Big(\sum_{1\le h_1<h_2\le N} e^{h_1}\wedge e^{h_2} \Big) \Big(\sum_{k=1}^{N} \delta^i_k e_k \Big) \wedge e_j .$$

Using an obvious short-hand notation the duality relation (A.30) can be compactly written as

$$\varepsilon^I_J = e^I(e_J) = \int \Big(\sum_H e^H \Big) \Big(\sum_K \varepsilon^I_K e_K \Big) \wedge e_J . \tag{A.31}$$

Next we introduce the space of Grassmann algebra valued test functions with support in $\mathbb{R}^d$:

$$\mathcal{S}(\mathbb{R}^d) \otimes G(V) .$$

An element of this space has the form

$$F(x; e_1, \dots, e_N) = \sum_{I=1}^{2^N} F^I(x)\, e_I \tag{A.32}$$

$$= \sum_{m=0}^{N} \sum_{1\le i_1<\dots<i_m\le N} F^{i_1\dots i_m}(x)\, e_{i_1}\wedge\dots\wedge e_{i_m}$$

$$= F^0(x) + \sum_{i=1}^{N} F^i(x)\, e_i + \dots + F^{1\dots N} e_1\wedge\dots\wedge e_N ,$$

where the $F^{i1..in}(x)$ are strongly decreasing, complex valued functions on $\mathbb{R}^d$. This expression coincides with the θ-expansion (A.27) apart from the fact that all component fields $F^{i1..in}$ in (A.32) are ordinary commuting functions.

To cure this problem we proceed as in the previous sections and introduce the <u>graded symmetric tensor algebra constructed from the space $\mathcal{S}(\mathbb{R}^d) \otimes G(V)$.</u> Indeed $\mathcal{S}(\mathbb{R}^d) \otimes G(V)$ admits a natural algebra structure defined by point-wise multiplication,

$$(F \cdot \tilde{F})(x; e_i) := F(x, e_i) \wedge \tilde{F}(x, e_i)$$

and inherits a $\mathbb{Z}_2$-graduation from G(V) (deg $\mathcal{S}(\mathbb{R}^d) \equiv 0$). Thus we can consider the graded tensor product of the graded algebra $\mathcal{S}(\mathbb{R}^d) \otimes G(V)$ with itself[191)] and more generally the tensor algebra

$$\mathcal{T}_{graded} = \bigoplus_{n=0}^{\infty} \underbrace{\left[\mathcal{T}(\mathbb{R}^d) \otimes G(V) \right]^{\otimes n}}_{\text{n-fold graded tensor product}} .$$

Typical elements of this space are sequences

$$F = \begin{bmatrix} F_0 \\ F_1(x, e_1, \dots, e_N) \\ F_2(x_1, e_1^{(1)}, \dots, e_N^{(1)}; x_2, e_1^{(2)}, \dots, e_N^{(2)}) \\ \vdots \end{bmatrix}$$

where the generators of the n-th copy of G(V) are supplemented with an upper index n: $e_1^{(n)}, \dots, e_N^{(n)}$.

Now let

$$\underline{\phi}(x; e^1, \dots, e^N) = \sum_{I=1}^{2^N} \underline{\phi}_I(x)\, e^I$$

denote an (improper) <u>generator in dimension one of the dual tensor algebra</u>

$$\left(\mathcal{T}_{graded}\right)^* \cong \bigoplus_{n=0}^{\infty} \left[\mathcal{T}'(\mathbb{R}^d) \otimes G(V^*) \right]^{\otimes n} .$$

Its action on elements $F \in \mathcal{T}_{graded}$ is given by

$$\underline{\phi}_I(x)\,[F] = \sum_J \varepsilon_I^J\, F_1^J(x) \qquad \text{(A.33a)}$$

i.e.

$$\underline{\phi}(x; e^i)[F] = \Big(\sum_I \underline{\phi}_I(x)\, e^I\Big)[F] = \Big(\sum_{I,J} \varepsilon_I^J\, F_1^J(x)\Big) e^I$$

$$\equiv F_1(x; e^i) \qquad \text{(A.33b)}$$

and more generally

$$\left(\underline{\phi}(x_1; e^{i_1}_{(1)}) \otimes \ldots \otimes \underline{\phi}(x_n; e^{i_n}_{(n)})\right)[F] = F_n\left(x_1, e^{i_1}_{(1)}; \ldots; x_n, e^{i_n}_{(n)}\right) . \tag{A.33c}$$

Here $e^{i_n}_{(n)}$ stands for $e^{1}_{(n)},\ldots,e^{N}_{(n)}$ and F_n is graded symmetric in all its arguments.

As pointed out before a <u>superfield functional</u> T_{gr} corresponds to a sequence of distribution-valued kernels, $T_{gr} \in (\mathcal{J}_{graded})^*$, and can be expressed as

$$T_{graded} = \sum_{n=0}^{\infty} \int dx_1 \int \sum_{H_1} e^{H_1}_{(1)} \ \ldots \ \int dx_n \int \sum_{H_n} e^{H_n}_{(n)}$$

$$\cdot\, T_n\left(x_1, e^{(1)}_{i_1}; \ldots; x_n, e^{(n)}_{i_n}\right) \underline{\phi}(x_1, e^{(1)}_{i_1}) \otimes \ldots \otimes \underline{\phi}(x_n, e^{(n)}_{i_n})$$

$$= 1 + \int dx \int \sum_{H} e^{H} \ T_1(x, e_i)\, \underline{\phi}(x, e_i) + \ldots$$

where the Grassmann algebra indices in T_n and $\underline{\phi}$ have been lowered according to eqs. (A.31), (A.33). Application to $F \in \mathcal{J}_{graded}$ yields

$$T_{graded}[F] = F_0 + \int dx \int \sum_{H} e^{H} \ T_1(x, e_i)\ \underline{\phi}(x, e_i)[F] + \ldots$$

$$= F_0 + \int dx \int \sum_{H} e^{H} \ T_1(x, e_i)\ F_1(x, e_i) + \ldots$$

$$+ \int dx_1 \int \sum_{H_1} e^{H_1}_{(1)} \ldots \int dx_n \int \sum_{H_n} e^{H_n}_{(n)} \ T_n\left(x_1, e^{(1)}_{i_1}; \ldots; x_n, e^{(n)}_{i_n}\right) \cdot$$

$$\cdot\, F_n\left(x_1, e^{(1)}_{i_1}; \ldots; x_n, e^{(n)}_{i_n}\right) + \ldots \tag{A.34}$$

In particular for the test functions

$$F = \begin{bmatrix} 1 \\ \phi \\ \phi \otimes \phi \\ \cdot \\ \cdot \\ \cdot \end{bmatrix}$$

we can recover the functional (A.28).

(v) Global aspects

The coupling to gravity requires the introduction of non-flat space-time manifolds. In the following we indicate how the framework presented so far can be generalized to this case. We only deal with the supersymmetric theory, the reasoning in the other cases going along the same lines. Our considerations are based on the Berezin-Leĭtes-Kostant-Batchelor approach to supermanifolds which we now summarize in a few words[195),196),88),197),214)].

Let M denote a real d-dimensional C^{∞}-manifold which admits a spin structure. Furthermore let E be a smooth, complex, rank N vector bundle over M with typical fibre V. Using the transition functions of E one can construct the Grassmann bundle $\Lambda E = \bigoplus_{p=0}^{N} \Lambda^p E$ over M; its typical fibre is given by the vector space

$$G(V) \equiv \Lambda V = \bigoplus_{p=0}^{N} \Lambda^p V \quad . \tag{A.35}$$

We observe that the linear space $\Gamma(\Lambda E)$ of smooth sections in ΛE has a natural Grassmann algebra structure induced by fiberwise exterior multiplication over each point $x \in M$.

Up to diffeomorphisms the base manifold M is determined by the sheaf of smooth functions on M (i.e. sections in the trivial vector bundle $M \times \mathbb{C}$ over M) *). To each open subset U of M this sheaf associates the commutative algebra $C^{\infty}(U)$:

---

*) The underlying idea comes from Gelfand's theory of C^*-algebras[211)] which asserts that a compact topological space M is characterized by the *-algebra $C^0(M)$ of continuous complex valued functions on M and that such algebras are the most general commutative C^*-algebras. These results are at the origin of the sheaf-theoretical definition of manifolds, see ref. 212).

$$M \supset \mathcal{U} \longmapsto C^{\infty}(\mathcal{U})$$

B. Kostant has generalized this idea to <u>supermanifolds</u> ("graded manifolds") and M. Batchelor has shown that in that case $C^{\infty}(U)$ has to be replaced by the space of local sections in $\wedge E$ (where E is some <u>non-canonically defined</u> vector bundle over an ordinary manifold M). (It is in this form that the supermanifold-definition had been given earlier in the Russian literature by F.A. Berezin and D.A. Leĭtes[195]).

To an open subset $U \subset M$ this sheaf associates the graded commutative algebra $\Gamma(\wedge E|U)$ of smooth sections in $\wedge E$ over U:

$$M \supset \mathcal{U} \longmapsto \Gamma(\wedge E|\mathcal{U}) .$$

A local section[213] in this sheaf has the form (A.32) with smooth component fields $F^{i_1 \ldots i_n}$. Thus all previous relations apply locally on M. If we consider open subsets $U \subset M$ which are associated with a local trivialization of the bundle E (and thereby of $\wedge E$), the local results can be patched together in a smooth way to yield a global description on M.

(vi) Concluding remarks

In the foregoing sections we have made a modest attempt to describe anti-commuting fermionic fields (and functionals thereof) while avoiding any ad hoc introduction of infinite dimensional Grassmann algebras that are not generated by the space-time structure. The generating functional Z[J] of (connected) Green's functions can be included in this framework by the consideration of an infinite dimensional <u>source algebra</u> which is constructed in analogy with and isomorphic to the field algebra. For their part the <u>n-point functions</u> are obtained by functionally differentiating with respect to the algebra generators $\underline{J}(x)$, $\underline{\varphi}(x)$, $\underline{\Psi}_i(x)$ and $\underline{\phi}(x,e_i)$ respectively.

REFERENCES

1) L.F. Abbott, M.T. Grisaru and D. Zanon, Nucl. Phys. B244 (1984) 454.

2) E. Abdalla, M. Forger and M. Jacques, "Higher Conservation Laws for Ten-Dimensional Supersymmetric Yang-Mills Theories", preprint CERN-TH.4431/86 (April 1986).

3) S. Adler, Phys. Rev. 177 (1969) 2426.

4) S. Adler and W. Bardeen, Phys. Rev. 182 (1969) 1517.

5) S. Adler, in "Lectures on Elementary Particles and Quantum Field Theory", Brandeis 1970, S. Deser et al. eds (MIT Press, 1970).

6) L. Alvarez-Gaumé, in "Fundamental Problems in Gauge Field Theory", Erice 1985, G. Velo and A.S. Wightman eds, NATO ASI Ser. B, vol. 141 (Plenum, 1986).

7) J.-P. Antoine and M. Jacques, Class. Quant. Grav. 1 (1984) 431 and Class. Quant. Grav. 4 (1987) 181.

8) R. Arnowitt, P. Nath and B. Zumino, Phys. Lett. 56 (1975) 81.

9) R. Arnowitt and P. Nath, Phys. Rev. D18 (1978) 2759.

10) R. Arnowitt and P. Nath, Nucl. Phys. B165 (1980) 462.

11) R. Arnowitt and P. Nath, in "Unification of the Fundamental Particle Interactions", Erice 1980, S. Ferrara, J. Ellis and P. van Nieuwenhuizen eds (Plenum, 1980).

12) M.F. Atiyah and I.M. Singer, Proc. Nat. Acad. Sci. USA 81 (1984) 2597.

13) W. Bardeen, Phys. Rev. 184 (1969) 1848.

14) L. Baulieu and J. Thierry-Mieg, Nucl. Phys. B197 (1982) 477.

15) L. Baulieu, Phys. Rep. 129C (1985) 1.

16) L. Baulieu and M. Bellon, Nucl. Phys. B266 (1986) 75.

17) L. Baulieu, B. Grossmann and R. Stora, Phys. Lett. B180 (1987) 95.

18) L. Baulieu, M. Bellon and R. Grimm, "BRS Symmetry of Supergravity in Superspace and its Projection to Component Formalism", preprint LAPP-TH-181, Dec. 1986.

19a) C. Becchi, A. Rouet and R. Stora, Phys. Lett. 52B (1974) 344 and Ann. Phys. 98 (1976) 287.

19b) C. Becchi, A. Rouet and R. Stora, in "Field Theory, Quantization and Statistical Physics", E. Tirapegui ed. (Reidel, 1981).

20) C. Becchi, in "Renormalization Theory", Erice 1975, G. Velo and A.S. Wightman eds, NATO ASI Ser. C, vol. 23 (Reidel, 1976) and in "Relativity, Groups and Topology II", Les Houches 1983, B. de Witt and R. Stora eds (North-Holland, 1984).

21) D. Bleecker, "Gauge Theory and Variational Principles" (Addison-Wesley, 1981).

22) N.N. Bogolubov, A.A. Logunov and I.T. Todorov, "Introduction to Axiomatic Quantum Field Theory" (Benjamin/Cummings, 1975).

23) P.J.M. Bongaarts, in "Mathematical Aspects of Gravity and Supergravity", NATO Advanced Workshop, July 1986, Logan, Utah, U.S.A. (to appear).

24) L. Bonora, P. Pasti and M. Tonin, Nuov. Cim. 63A (1981) 353.

25) L. Bonora and P. Cotta-Ramusino, Comm. Math. Phys. 87 (1983) 589.

26) L. Bonora, P. Pasti and M. Tonin, Phys. Lett. 156B (1985) 341; Nucl. Phys. B261 (1985) 249, (see also Errata in Nucl. Phys. B269 (1986) 745).

27) L. Bonora, P. Pasti and M. Tonin, "Comment on the Supersymmetry Anomaly in the WZ-Gauge", (unpublished).

28) M. Breitenecker and H.-R. Grümm, Nuov. Cim. 55A (1980) 453.

29) P. Breitenlohner and D. Maison, in "Supersymmetry and its Applications : Superstrings, Anomalies and Supergravity", Nuffield Workshop, 1985, G.W. Gibbons, S.W. Hawking and P.K. Townsend eds (Cambridge University Press, 1986).

30) L. Brink, J.H. Schwarz and J. Scherk, Nucl. Phys. B121 (1977) 77.

31) U. Bruzzo and R. Cianci, Class. Quant. Grav. 1 (1984) 213.

32) J. Buckingham and A.W. Fisher, "Coupled Chiral and Supersymmetry Anomalies in Supersymmetric Yang-Mills Theories", preprint King's College, London, Sept. 1986.

33) S.S. Chang, Phys. Rev. D14 (1976) 446.

34) T.-P. Cheng and L.-F. Li, "Gauge Theory of Elementary Particle Physics" (Clarendon Press, Oxford 1984).

35) S.-S. Chern, "Complex Manifolds without Potential Theory" (Springer 1979).

36) Y. Choquet-Bruhat, C. de Witt-Morette with M. Dillard-Bleick, "Analysis, Manifolds and Physics" (North-Holland, 1982).

37) C. Curci and R. Ferrari, Phys. Lett. 63B (1976) 91.

38) B. Delamotte and F. Delduc, Phys. Lett. 182B (1986) 337.

39) B. Delamotte and J. Kaplan, Class. Quant. Grav. 4 (1987) 1223 and B. Delamotte, Thesis (Univ. Paris VII, Sept. 1987), unpublished.

40) B. de Wit and D.Z. Freedman, Phys. Rev. D12 (1975) 2286.

41) B. de Witt, "Dynamical Theory of Groups and Fields", (Gordon & Breach, 1965).

42) B. de Witt, "Supermanifolds" (Cambridge University Press, 1984).

43) M. Dubois-Violette, M. Talon and C.M. Viallet, Phys. Lett. 158B (1985) 231; Comm. Math. Phys. 101 (1985).

44) M. Dubois-Violette, in "Fields and Geometry", Karpacz 1986, A. Jadcyzk ed. (World Scientific, 1986).

45) T. Eguchi, P.B. Gilkey and A.J. Hanson, Phys. Rep. 66 (1980) 213.

46) L.D. Fadeev and V.N. Popov, Phys. Lett. 25B (1967) 29.

47) L.D. Fadeev and A.A. Slavnov, "Gauge Fields : Introduction to Quantum Theory", (Benjamin/Cummings, 1980).

48) L.D. Fadeev, Phys. Lett. 145B (1984) 81.
L.D. Fadeev and S. Shatashvili, Phys. Lett. 167B (1986) 225.

49) P. Fayet, Nucl. Phys. B113 (1976) 135.

50) S. Ferrara, B. Zumino and J. Wess, Phys. Lett. 51B (1974) 239.

51) S. Ferrara and B. Zumino, Nucl. Phys. B79 (1974) 413.

52) S. Ferrara, L. Girardello, O. Piguet and R. Stora, Phys. Lett. 157B (1985) 179.

53) R. Feynman, Act. Phys. Pol. XXIV (1963) 697.

54) P.G.O Freund, "Introduction to Supersymmetry" (Cambridge Univ. Press, 1986).

55) A. Galperin, E. Ivanov, S. Kalitzin, V. Ogievetsky and E. Sokatchev, Class. Quant. Grav. 1 (1984) 469, "Unconstrained off-shell N=3 Supersymmetric Yang-Mills Theory", preprint E2-84-441 (1984), subm. to Class. Quant. Grav. and in "Supersymmetry, Supergravity, Superstrings 1986", Trieste 1986, B. de Witt, P. Fayet and M. Grisaru eds (World Scientific 1986).

56) R. Garreis, M. Scholl and J. Wess, Z. Phys. C28 (1985) 623;
R. Garreis and M. Scholl in "Fields and Geometry", Karpacz 1986, A. Jadczyk ed. (World Scientific 1986).

57) S.J. Gates, M.T. Grisaru, M. Roček and W. Siegel, "Superspace" (Benjamin/ Cummings, 1983).

58) S.J. Gates Jr., K.S. Stelle and P.C. West, Nucl. Phys. B169 (1980) 347.

59) F. Gieres, "Über den Aharonov-Bohm-Effekt", Diploma Thesis, Univ. of Göttingen, 1983 (unpublished).

60) F. Gieres, "Geometric Approach to the (BRS-) Differential Algebras of Supersymmetric YM-Theories", preprint LAPP-TH-184/87, Jan. 1987 (to be published in Class. Quant. Grav.).

61) R. Gilmore, "Lie Groups, Lie Algebras and some of their Representations", (John Wiley & Sons, Inc., 1974).

62) G. Girardi, R. Grimm and R. Stora, Phys. Lett. 156B (1985) 203.

63) W. Greub, S. Halpern and R. Vanstone, "Connections, Curvature and Cohomology", Vol. III (Academic Press, 1976).

64) P.A. Griffiths and J.W. Morgan, "Rational Homotopy Theory and Differential Forms", Progress in Math. Vol. 16 (Birkhäuser, 1981).

65) R. Grimm, M. Sohnius and J. Wess, Nucl. Phys. B133 (1978) 275.

66) R. Grimm, in "Mathematical Aspects of Gravity and Supergravity", NATO Advanced Workshop, July 1986, Logan, Utah, U.S.A. (to appear).

67) J. Grinberg, Helv. Phys. Act. 57 (1984) 321.

68) E. Guadagnini, K. Konishi and M. Mintchev, Phys. Lett. 157B (1985) 37.

69) E. Guadagnini and M. Mintchev, Nucl. Phys. B269 (1986) 543.

70a) A. Guichardet, "Cohomologie des Groupes Topologiques et des Algèbres de Lie" (Cedic-Nathan, Paris, 1980).

70b) D.B. Fuks, "Cohomology of Infinite-Dimensional Lie Algebras", (Consultants Bureau/Plenum Publ. Comp., 1986).

71) J. Harnad, J. Hurtubise, M. Légaré and S. Shnider, Nucl. Phys. B256 (1985) 609.

72) J. Harnad and S. Shnider, Comm. Math. Phys. 106 (1986) 183.

73) S. Helgason, "Differential Geometry, Lie Groups and Symmetric Spaces" (Academic Press, 1978).

74) H. Holmann and H. Rummler, "Alternierende Differentialformen" (B.I., 1981).

75) P.S. Howe, K.S. Stelle and P.K. Townsend, Nucl. Phys. B236 (1984) 125.

76) D.S. Hwang, Nucl. Phys. B267 (1986) 349.

77) M. Ito, T. Morozumi, H. Nojiri and J. Uehara, "Covariant Quantization of Neveu-Schwarz-Ramond Model", preprint RRK 85-34, Nov. 1985.

78) H. Itoyama, V.P. Nair and H. Ren, Nucl. Phys. B262 (1985) 317 and Phys. Lett. 168B (1986) 78.

79) C. Itzykson and J.-B. Zuber, "Quantum Field Theory" (McGraw-Hill, 1980).

80) R. Jackiw and K. Johnson, Phys. Rev. 182 (1969) 1459.

81) R. Jackiw, in "Lectures on Current Algebra and its Applications", S.B. Treiman, R. Jackiw and D.J. Gross eds (Princeton University Press, 1972).

82) A. Jadczyk and K. Pilch, Comm. Math. Phys. 78 (1981) 373.

83) L. Kannenberg, J. Math. Phys. 19 (1978) 2203.

84) D. Kastler and R. Stora, "A Differential Geometric Setting for BRS Transformations and Anomalies I, II", preprints CPT-Marseille and LAPP-Annecy-le-Vieux.

85) S. Kobayashi and K. Nomizu, "Foundations of Differential Geometry I & II" (John Wiley & Sons, 1963 & 1969).

86) Y. Kobayashi and S. Nagamachi, Lett. Math. Phys. 11 (1986) 293.

87) J. Koller, Nucl. Phys. B222 (1983) 319 and Phys. Lett. 124B (1983) 324.

88) B. Kostant, in "Differential Geometric Methods in Mathematical Physics", Lect. Notes in Math., vol. 570 (Springer, 1977).

89) T. Kugo and I. Ojima, Suppl. Progr. Theor. Phys. 66 (1979) 1.

90) W. Kummer and M. Schweda, Phys. Lett. 141B (1984) 363;
W. Kummer, T. Kreutzberger, O. Piguet, A. Rebhan and M. Schweda, Phys. Lett. 167B (1986) 393.
W. Kummer, H. Mistelberger, P. Schaller, M. Schweda and T. Kreutzberger, Nucl. Phys. B281 (1987) 411.

91) F. Langouche, T. Schücker and R. Stora, Phys. Lett. 145B (1984) 342.

92) B. Lautrup, Mat. Fys. Medd. Dan. Vid. Selsk 35 (1967) Nr.11.

93) J.M. Leinaas and K. Olaussen, Phys. Lett. 108B (1982) 199.

94) J. Mañes, R. Stora and B. Zumino, Comm. Math. Phys. 102 (1985) 157.

95) J. Martin-Hernandez and J.G. Taylor, Phys. Lett. B185 (1987) 99.

96) M.E. Mayer, in "New Developments in Mathematical Physics", Acta Phys. Austriaca Suppl. XXIII (1981) 491 (Springer 1981) and Phys. Lett. 120B (1983) 355.

97) I.N. McArthur and H. Osborn, Nucl. Phys. B268 (1986) 573.

98) L. Mezincescu, Dubna JINR Report P2-12572 (1979).

99) R. Musto, L. O'Raifeartaigh and A. Wipf, Phys. Lett. 175B (1986) 433.

100) N. Nakanishi, Progr. Theor. Phys. 35 (1966) 1111.

101) C. Nash and S. Sen, "Topology and Geometry for Physicists" (Academic Press 1983).

102) Y. Ne'eman and T. Regge, Riv. Nuov. Cim. 1 (1978) 1.

103) N.K. Nielsen, H. Römer and B. Schroer, Phys. Lett. 70B (1977) 445; N.K. Nielsen and B. Schroer, Nucl. Phys. B127 (1977) 493.

104) N.K. Nielsen, Nucl. Phys. B244 (1984) 499 and in "Topological and Geometrical Methods in Field Theory", Espoo 1986, J. Hietarinta and J. Westerholm eds (World Scientific 1986).

105) N. Ohta, Phys. Rev. D33 (1986) 1681.

106) I. Ojima, Progr. Theor. Phys. 64 (1980) 625.

107) S. Olariu and I.I. Popescu, Rev. Mod. Phys. 57 (1985) 339.

108) F.R. Ore Jr. and P. van Nieuwenhuizen, Nucl. Phys. B204 (1982) 317.

109) R. Percacci, "Geometry of non-linear Field Theories", (World Scientific 1986).

110) M. Pernici and F. Riva, Nucl. Phys. B267 (1986) 61.

111) O. Piguet and A. Rouet, Phys. Rep. 76 (1981) 1.

112) O. Piguet, M. Schweda and K. Sibold, Nucl. Phys. B174 (1980) 183.

113) O. Piguet, in "New Trends in Particle Theory", John Hopkins Workshop, 1985, L. Lusanna ed. (World Scientific, 1985).

114) O. Piguet and K. Sibold, "Renormalized Supersymmetry. The Perturbation Theory of N=1 Supersymmetric Theories in Flat Space-Time" (Birkhäuser, 1986).

115) T. Regge, in "Relativity, Groups and Topology II", B. de Witt and R. Stora eds (North-Holland 1984).

116) A. Rogers, J. Math. Phys. 21 (1980) 1352 and J. Math. Phys. 22 (1981) 939.

117) A. Rogers, Comm. Math. Phys. 105 (1986) 375.

118) A. Salam and J. Strathdee, Nucl. Phys. B76 (1974) 477 and Phys. Rev. D11 (1975) 1521.

119) A. Salam and J. Strathdee, Nucl. Phys. B80 (1974) 499.

120) T. Schücker, preprint HD-THEP-86-5 (submitted to Comm. Math. Phys.).

121) A.S. Schwarz and A.A. Rosly, Comm. Math. Phys. 105 (1986) 645.

122) H.-J. Seifert, C.J.S. Clarke and A. Rosenblum (eds), "Mathematical Aspects of Superspace", Hamburg 1983, NATO ASI Ser. C, Vol. 132 (D. Reidel Publ. 1984).

123) D. Sen, J. Math. Phys. 27 (1986) 472 and Nucl. Phys. B284 (1987) 201.

124) J. Sidenius, "Anomalies and Differential Geometry", preprint NORDITA-85/33 (March 1985).

125) I.M. Singer, in Colloque Elie Cartan, Lyon, June 25-29, 1984 (Astérisque, 1985).

126) M. Sohnius, Nucl. Phys. B136 (1978) 461.

127) M. Sohnius, K.S. Stelle and P.C. West, in "Superspace and Supergravity", Nuffield Workshop 1980, S.W. Hawking and M. Roček eds (Cambridge Univ. Press, 1981).

128) M. Sohnius, Phys. Rep. 128 (1985) 39.

129) P.P. Srivastava, Lett. Nuov. Cim. 13 (1975) 657.

130) P.P. Srivastava, "Supersymmetry, Superfields and Supergravity: an Introduction" (Adam Hilger, 1986).

131) J. Stasheff, in "Symposium on Anomalies, Geometry, Topology", W.A. Bardeen and A.R. White eds (World Scientific, 1985).

132) K.S. Stelle and P. West, Nucl. Phys. B140 (1978) 285.

133) R. Stora, in "Renormalization Theory", Erice 1985, G. Velo and A.S. Wightman eds, NATO ASI Ser. C, Vol. 23 (Reidel, 1976).

134) R. Stora, in "New Developments in Quantum Field Theory and Statistical Mechanics", Cargèse 1976, M. Lévy, P. Mitter eds, NATO ASI Ser. B, Vol. 26 (Plenum Press, 1977).

135) R. Stora, in "Progress in Gauge Field Theory", Cargèse 1983, G. 't Hooft et al. eds, NATO ASI Ser. B, Vol. 115 (Plenum Press, 1984).

136) R. Stora, in "New Perspectives in Quantum Field Theory", J. Abat, M. Asorey and A. Cruz eds (World Scientific, 1986).

137) E.C.G. Stueckelberg, Helv. Phys. Acta 11 (1938) 225, 299.

138) M. Talon, "Algebra of Anomalies", Lectures at 1985 Cargèse Summer School (Plenum).

139) M. Talon, "BRS Algebra and Anomalies", preprint LPTHE 86.38, Paris, Sept. 1986.

140) Y. Tanii, Nucl. Phys. B289 (1987) 187.

141) J. Thierry-Mieg, Nuov. Cim. 56A (1980) 396.

142) J. Thierry-Mieg and Y. Ne'eman, in Proceedings of the "Second Marcel Grossmann Meeting on General Relativity", R. Ruffini ed. (North-Holland Publ. Co. 1982).

143) W. Thirring, "Classical Field Theory. A Course in Mathematical Physics II" (Springer, 1979).

144) I.V. Tyutin, Report Fian 39 (1975) (unpublished).

145) P. van Nieuwenhuizen, Phys. Rep. 68 (1981) 189.

146) P. van Nieuwenhuizen, in "Supersymmetry and Supergravity", Trieste 1984, B. de Witt, P. Fayet and P. van Nieuwenhuizen eds (World Scientific, 1984).

147) C. Viallet, in "Fields and Geometry", Karpacz 1986, A. Jadcyzk ed. (World Scientific, 1986).

148) H.C. Wang, Nagoya Math. J. 13 (1958) 1.

149) Zi Wang and Yong-Shi Wu, Phys. Lett. 164B (1985) 305.

150) F.W. Warner, "Foundations of Differential Manifolds and Lie Groups" (Springer 1983).

151) J. Wess and B. Zumino, Phys. Lett. 37B (1971) 95.

152) J. Wess and B. Zumino, Nucl. Phys. B78 (1974) 1.

153) J. Wess, in "Topics in Quantum Field Theory", J.A. Azcárraga ed., Lect. Notes in Phys. 77 (Springer, 1978).

154) J. Wess and J. Bagger, "Supersymmetry and Supergravity", (Princeton University Press, 1983).

155) P. West, "Introduction to Supersymmetry and Supergravity", (World Scientific, 1986).

156) P. West, "Supersymmetry: A Decade of Development", P.West ed. (Adam Hilger,1986).

157) A.S. Wightman, in "Relations de Dispersion et Particules Elémentaires", Les Houches 1960, C. de Witt and R. Omnès eds (Hermann, Paris 1960).

158) E. Witten, Phys. Lett. 77B (1978) 394.

159) G. Woo, Lett. Nuov. Cim. 13 (1975) 546.

160) T.T. Wu and C.N. Yang, Phys. Rev. D12 (1975) 3845.

161) J. Zinn-Justin, in "Trends in Elementary Particle Theory", Bonn 1974, K. Dietz and H. Rollnik eds, Lecture Notes in Physics Vol. 37 (Springer, 1975).

162) B. Zumino, in "Recent Developments in Gravitation", Cargèse 1978, M. Lévy and S. Deser eds, NATO ASI Ser. B, vol. 44 (Plenum Press, 1979).

163) B. Zumino, in "Unification of the Fundamental Particle Interactions", Erice 1980, S. Ferrara, J. Ellis and P. van Nieuwenhuizen eds (Plenum, 1980).

164) B. Zumino, in "Relativity, Groups, Topology II", Les Houches 1983, B.S. de Witt and R. Stora eds (North-Holland Publ., Amsterdam 1984).

165) B. Zumino, in "Symposium on Anomalies, Geometry, Topology", W.A. Bardeen and A.R. White eds (World Scientific, 1985).

166) Y. Aharonov and D. Bohm, Phys. Rev. 115 (1959) 485.

167) D. Sullivan, IHES Pub. N° 47 (1977) 269.

168) P.K. Townsend and P. van Nieuwenhuizen, Nucl. Phys. B120 (1977) 301.

169) L. Baulieu and J. Thierry-Mieg, Phys. Lett. 145B (1984) 53.

170) L. Baulieu, C. Becchi and R. Stora, Phys. Lett. 180B (1986) 55 ;
C. Becchi, "On the Covariant Quantization of the Free String : the Conformal Structure", preprint University of Genova (June 1987), and
R. Stora, "Algèbres Différentielles en Théorie des Champs", Colloque International de Géométrie en l'honneur de J.L. Koszul, Grenoble, 1-6 Juin 1987, preprint LAPP-TH-195/87 (June 1987).

171) G. Sterman, P.K. Townsend and P. van Nieuwenhuizen, Phys. Rev. D17 (1978) 1501.

172) H. Hata and T. Kugo, Nucl. Phys. B158 (1979) 357.

173) M.B. Green, J.H. Schwarz and E. Witten, "Superstring Theory I,II" (Cambridge University Press, 1987).

174) J. Glimm and A. Jaffe, "Quantum Physics - A Functional Integral Point of View" (Springer, 1981).

175a) J. Steinberger, Phys. Rev. 76 (1949) 1180.

175b) J. Schwinger, Phys. Rev. 82 (1951) 664.

175c) J.S. Bell and R. Jackiw, Nuov. Cim. 60A (1969) 47.

176) B. Schroer, in "Facts and Prospects of Gauge Theories", Act. Phys. Austr. Suppl. XIX, Schladming 1978 (Springer 1978).

177) F. Gliozzi, J. Scherk and D. Olive, Nucl. Phys. B122 (1977) 253.

178) M. Müller, Nucl. Phys. B289 (1987) 557.

179) M. Batchelor, Trans. Amer. Math. Soc. Vol. 253 (1979) 329 and Vol. 258 (1980) 257 ; see also lectures in ref. 122).

180) J. Dell and L. Smolin, Comm. Math. Phys. 66 (1979) 197.

181) V.G. Kac, Adv. in Math. 26 (1977) 8.

182) R. Stora, "Superfields and Functionals of Superfields", Unpublished Notes (July 1985)

183) B. Zumino, Wu Yong-Shi and A. Zee, Nucl. Phys. B239 (1984) 477 and B. Zumino, Nucl. Phys. B253 (1985) 477.

184) A.W. Fisher, Nucl. Phys. B287 (1987) 144.

185) J.-L. Koszul, Bull. Soc. Math. France 78 (1950) 65.

186) T. Kawai, Phys. Lett. 168B (1986) 355.

187) D. Altschüler, Mod. Phys. Lett. A1 (1986) 557.

188) I.B. Frenkel, H. Garland and G.J. Zuckerman, "Semi-infinite Cohomology and String Theory", preprint Dept. of Math., Yale University (1986).

189) M. Karoubi, "K-Theory. An Introduction" (Springer, 1978).

190) M. Henneaux, Phys. Rep. 126 (1985) 1.

191) M. Scheunert, "The Theory of Lie Superalgebras. An Introduction", Lecture Notes in Mathematics 716 (Springer, 1979).

192) C. Chevalley and S. Eilenberg, Trans. Am. Math. Soc. 63 (1948) 85.

193) L. Bonora and M. Tonin, Phys. Lett. 98B (1981) 48.

194) C.P. Luehr and M. Rosenbaum, J. Math. Phys. 28 (1987) 2053.

195) F.A. Berezin and D.A. Leĭtes, Soviet Math. Dokl. 16 (1975) 1218;
F.A. Berezin, Sov. J. Nucl. Phys. 29 (1979) 857;
D.A. Leĭtes, Russ. Math. Surv. 35 (1980) 1.

196) F.A. Berezin, "Introduction to Superanalysis", ed. A.A. Kirillov, Math. Phys. and Appl. Math. 9 (D. Reidel Publ., 1986).

197) M. Batchelor, Trans Amer. Math. Soc. Vol.253 (1979) 329; ibidem Vol. 258 (1980) 257 and review article in ref. 122).

198) A. Rogers, Comm. Math. Phys. 113 (1987) 353.

199) "Selected Papers (1937-1976) of Julian Schwinger", eds. M. Flato, C. Fronsdal and K.A. Milton (D. Reidel Publ., 1979).

200) P.T. Mathews and A. Salam, Nuov. Cim. 12 (1954) 563 and ibid. 2 (1955) 120.

201) F.A. Berezin: "The Method of Second Quantization" (Academic Press, 1966).

202) K. Osterwalder and R. Schrader, Phys. Rev. Lett. 29 (1972) 1423; Helv. Phys. Acta 46 (1973) 277 and Comm. Math. Phys. 31 (1973) 83.

203) F.A. Berezin, Sov. J. Nucl. Phys. 30 (1979) 605;
Y. Ne'eman, in "Differential Geometric Methods in Mathematical Physics", eds. H.D. Doebner and J.D. Hennig, Lecture Notes in Mathematics 1139 (Springer, 1985);
R.F. Picken and K. Sundermeyer, Comm. Math. Phys. 102 (1986) 585;
B. de Witt, "Supermanifolds" (Cambridge Univ. Press, 1984);
A. Rogers, J. Math. Phys. 26 (1985) 385, 2749;
J.M. Rabin, Comm. Math. Phys. 103 (1986) 431;
J. Muñoz Masqué and D.H. Ruipérez, in "Differential Geometric Methods in Mathematical Physics", Lecture Notes in Mathematics 1251 (Springer, 1987);
M.J. Rothstein, Trans. Am. Math. Soc. 299 (1987) 387.

204) L. Schwartz, "Théorie des distributions" (Hermann, Paris 1950).

205) See chap. I.1 of R. Jost, "The General Theory of Quantized Fields" (Am.

Math. Soc., Provicence, 1965).

206) F. Treves, "Topological Vector Spaces, Distributions and Kernels" (Academic Press, 1967).

207) e.g. see appendix A of P. Kristensen, L. Mejlbo and E. Thue Poulsen, Comm. Math. Phys 1 (1965) 175.

208) H.-J. Borchers, in "Statistical Mechanics and Field Theory", Haifa 1971, eds. R.N. Sen and C. Weil (John Wiley & Sons, 1972); J. Yngvason,"Axiomatische Quantenfeldtheorie I", Göttingen Course, 1981-82 (unpublished).

209) I.M. Gel'fand and N.Y. Vilenkin, "Generalized Functions", Vol. 4 (Academic Press, 1964).

210) A. Böhm, "The rigged Hilbert space and Quantum Mechanics", Lecture Notes in Physics 78 (Springer, 1978).

211) I.M. Gel'fand, D.A. Raikow and G.E. Schilow, "Kommutative Normierte Algebren", (VEB, Deutscher Verlag der Wissenschaften, Berlin, 1964); K. Maurin, "Methods of Hilbert Spaces" (PWN, Polish Scient. Publ., Warszawa, 1967).

212) B.R. Tennison, "Sheaf Theory",London Math. Soc. Lect. Note Ser. 20 (Cambridge Univ. Press, 1975).

213) R.O. Wells, "Differential analysis on Complex Manifolds", Graduate Texts in Math. 65 (Springer, 1980).

214) Yu. I. Manin, Russ.Math. Surv. 39 (1984) 51 and "Kalibrovochnye polya i golomorfnaya geometriya"(Nauka, Moscow 1984); English translation "Gauge Fields and Complex Geometry", to appear (Springer Verlag)

SUBJECT INDEX

The underlined references indicate definitions or detailed discussions.

Abelian gauge theory 63, 107

Action
-, induced 5
-, left 5ff, 9, 10, 24
-, right 9ff
- of the adjoint representation 4
- of the Lorentz group 8
Action 151
-, effective 86, 87, 90, 91
-, quantum 100
- for a YM-field 38, 86
- for SYM 38ff, 47, 60, 68, 82ff
(see also ghost -)
Adiabatic limit 114
Affine
- motion 21
(see also - connection)
Aharonov-Bohm effect 84ff
-, supersymmetric 85
Algebra
-, * 165
-, abelian 3
-, bigraded 94
-, Borchers 156
-, C* 165
-, Clifford 146
-, commutative 1, 165
-, component field 90, 114, 115, 125, 127, 129
-, of covariant derivatives 50
-, de Rham 95
-, differential 86, 93, 108ff, 118ff, 137
-, Dirac 18, 150, 160
-, of (differential) forms 14, 15, 93
-, exterior 88, 93, 97, 101
-, field dependent 50, 86, 129
-, gauge 87, 89, 129
-, graded 3, 4, 32, 117, 162
-, graded commutative 1, 94, 166
-, Lorentz 30, 44
- morphism 25
-, of superfields 1
-, super Lie 3
-, super Poincaré 2, 5, 17, 25
-, supersymmetry 3, 4, 5, 16, 25, 48, 80ff, 111, 114, 130ff
-, Virasoro 149
(see also BRS-, cohomology-, Grassmann-, Lie-, tensor-, test function-algebra)
Anholonomy 25, 32
- coefficients 7, 11, 16, 17
Annihilation operator 147

Lecture Notes in Mathematics

Vol. 1174: Categories in Continuum Physics, Buffalo 1982. Seminar. Edited by F.W. Lawvere and S.H. Schanuel. V, 126 pages. 1986.

Vol. 1184: W. Arendt, A. Grabosch, G. Greiner, U. Groh, H.P. Lotz, U. Moustakas, R. Nagel, F. Neubrander, U. Schlotterbeck, One-parameter Semigroups of Positive Operators. Edited by R. Nagel. X, 460 pages. 1986.

Vol. 1186: Lyapunov Exponents. Proceedings, 1984. Edited by L. Arnold and V. Wihstutz. VI, 374 pages. 1986.

Vol. 1187: Y. Diers, Categories of Boolean Sheaves of Simple Algebras. VI, 168 pages. 1986.

Vol. 1190: Optimization and Related Fields. Proceedings, 1984. Edited by R. Conti, E. De Giorgi and F. Giannessi. VIII, 419 pages. 1986.

Vol. 1191: A.R. Its, V.Yu. Novokshenov, The Isomonodromic Deformation Method in the Theory of Painlevé Equations. IV, 313 pages. 1986.

Vol. 1194: Complex Analysis and Algebraic Geometry. Proceedings, 1985. Edited by H. Grauert. VI, 235 pages. 1986.

Vol. 1203: Stochastic Processes and Their Applications. Proceedings, 1985. Edited by K. Itô and T. Hida. VI, 222 pages. 1986.

Vol. 1209: Differential Geometry, Peñíscola 1985. Proceedings. Edited by A.M. Naveira, A. Ferrández and F. Mascaró. VIII, 306 pages. 1986.

Vol. 1214: Global Analysis – Studies and Applications II. Edited by Yu.G. Borisovich and Yu.E. Gliklikh. V, 275 pages. 1986.

Vol. 1218: Schrödinger Operators, Aarhus 1985. Seminar. Edited by E. Balslev. V, 222 pages. 1986.

Vol. 1227: H. Helson, The Spectral Theorem. VI, 104 pages. 1986.

Vol. 1229: O. Bratteli, Derivations, Dissipations and Group Actions on C*-algebras. IV, 277 pages. 1986.

Vol. 1236: Stochastic Partial Differential Equations and Applications. Proceedings, 1985. Edited by G. Da Prato and L. Tubaro. V, 257 pages. 1987.

Vol. 1237: Rational Approximation and its Applications in Mathematics and Physics. Proceedings, 1985. Edited by J. Gilewicz, M. Pindor and W. Siemaszko. XII, 350 pages. 1987.

Vol. 1250: Stochastic Processes – Mathematics and Physics II. Proceedings 1985. Edited by S. Albeverio, Ph. Blanchard and L. Streit. VI, 359 pages. 1987.

Vol. 1251: Differential Geometric Methods in Mathematical Physics. Proceedings, 1985. Edited by P.L. García and A. Pérez-Rendón. VII, 300 pages. 1987.

Vol. 1255: Differential Geometry and Differential Equations. Proceedings, 1985. Edited by C. Gu, M. Berger and R.L. Bryant. XII, 243 pages. 1987.

Vol. 1256: Pseudo-Differential Operators. Proceedings, 1986. Edited by H.O. Cordes, B. Gramsch and H. Widom. X, 479 pages. 1987.

Vol. 1258: J. Weidmann, Spectral Theory of Ordinary Differential Operators. VI, 303 pages. 1987.

Vol. 1260: N.H. Pavel, Nonlinear Evolution Operators and Semigroups. VI, 285 pages. 1987.

Vol. 1263: V.L. Hansen (Ed.), Differential Geometry. Proceedings, 1985. XI, 288 pages. 1987.

Vol. 1267: J. Lindenstrauss, V.D. Milman (Eds), Geometrical Aspects of Functional Analysis. Seminar. VII, 212 pages. 1987.

Vol. 1269: M. Shiota, Nash Manifolds. VI, 223 pages. 1987.

Vol. 1270: C. Carasso, P.-A. Raviart, D. Serre (Eds), Nonlinear Hyperbolic Problems. Proceedings, 1986. XV, 341 pages. 1987.

Vol. 1272: M.S. Livšic, L.L. Waksman, Commuting Nonselfadjoint Operators in Hilbert Space. III, 115 pages. 1987.

Vol. 1273: G.-M. Greuel, G. Trautmann (Eds), Singularities, Representation of Algebras, and Vector Bundles. Proceedings, 1985. XIV, 383 pages. 1987.

Lecture Notes in Physics

Vol. 281: Ph. Blanchard, Ph. Combe, W. Zheng, Mathematical and Physical Aspects of Stochastic Mechanics. VIII, 171 pages. 1987.

Vol. 282: F. Ehlotzky (Ed.), Fundamentals of Quantum Optics II. Proceedings, 1987. X, 289 pages. 1987.

Vol. 283: M. Yussouff (Ed.), Electronic Band Structure and Its Applications. Proceedings, 1986. VIII, 441 pages. 1987.

Vol. 284: D. Baeriswyl, M. Droz, A. Malaspinas, P. Martinoli (Eds.), Physics in Living Matter. Proceedings, 1986. V, 180 pages. 1987.

Vol. 285: T. Paszkiewicz (Ed.), Physics of Phonons. Proceedings, 1987. X, 486 pages. 1987.

Vol. 286: R. Alicki, K. Lendi, Quantum Dynamical Semigroups and Applications. VIII, 196 pages. 1987.

Vol. 287: W. Hillebrandt, R. Kuhfuß, E. Müller, J.W. Truran (Eds.), Nuclear Astrophysics. Proceedings. IX, 347 pages. 1987.

Vol. 288: J. Arbocz, M. Potier-Ferry, J. Singer, V.Tvergaard, Buckling and Post-Buckling. VII, 246 pages. 1987.

Vol. 289: N. Straumann, Klassische Mechanik. XV, 403 Seiten. 1987.

Vol. 290: K.T. Hecht, The Vector Coherent State Method and Its Application to Problems of Higher Symmetries. V, 154 pages. 1987.

Vol. 291: J.L. Linsky, R.E. Stencel (Eds.), Cool Stars, Stellar Systems, and the Sun. Proceedings, 1987. XIII, 537 pages. 1987.

Vol. 292: E.-H. Schröter, M. Schüssler (Eds.), Solar and Stellar Physics. Proceedings, 1987. V, 231 pages. 1987.

Vol. 293: Th. Dorfmüller, R. Pecora (Eds.), Rotational Dynamics of Small and Macromolecules. Proceedings, 1986. V, 249 pages. 1987.

Vol. 294: D. Berényi, G. Hock (Eds.), High-Energy Ion-Atom Collisions. Proceedings, 1987. VIII, 540 pages. 1988.

Vol. 295: P. Schmüser, Feynman-Graphen und Eichtheorien für Experimentalphysiker. VI, 217 Seiten. 1988.

Vol. 296: M. Month, S. Turner (Eds.), Frontiers of Particle Beams. XII, 700 pages. 1988.

Vol. 297: A. Lawrence (Ed.), Comets to Cosmology. X, 415 pages. 1988.

Vol. 298: M. Georgiev, F' Centers in Alkali Halides. XI, 282 pages. 1988.

Vol. 299: J.D. Buckmaster, T. Takeno (Eds.), Mathematical Modeling in Combustion Science. Proceedings, 1988. VI, 168 pages. 1988.

Vol. 300: B.-G. Englert, Semiclassical Theory of Atoms. VII, 401 pages. 1988.

Vol. 301: G. Ferenczi, F. Beleznay (Eds.), New Developments in Semiconductor Physics. Proceedings, 1988. VI, 308 pages. 1988.

Vol. 302: F. Gieres, Geometry of Supersymmetric Gauge Theories. VIII, 189 pages. 1988.